SpringerBriefs in Applied Sciences and Technology

SpringerBriefs present concise summaries of cutting-edge research and practical applications across a wide spectrum of fields. Featuring compact volumes of 50 to 125 pages, the series covers a range of content from professional to academic.

Typical publications can be:

- A timely report of state-of-the art methods
- An introduction to or a manual for the application of mathematical or computer techniques
- A bridge between new research results, as published in journal articles
- A snapshot of a hot or emerging topic
- An in-depth case study
- A presentation of core concepts that students must understand in order to make independent contributions

SpringerBriefs are characterized by fast, global electronic dissemination, standard publishing contracts, standardized manuscript preparation and formatting guidelines, and expedited production schedules.

On the one hand, **SpringerBriefs in Applied Sciences and Technology** are devoted to the publication of fundamentals and applications within the different classical engineering disciplines as well as in interdisciplinary fields that recently emerged between these areas. On the other hand, as the boundary separating fundamental research and applied technology is more and more dissolving, this series is particularly open to trans-disciplinary topics between fundamental science and engineering.

Indexed by EI-Compendex, SCOPUS and Springerlink.

More information about this series at http://www.springer.com/series/8884

Victor Bachinsky · Oleh Ya Vanchulyak ·
Alexander G. Ushenko · Yurii A. Ushenko ·
Alexander V. Dubolazov · Alexander Bykov ·
Benjamin Hogan · Igor Meglinski

Multi-parameter Mueller Matrix Microscopy for the Expert Assessment of Acute Myocardium Ischemia

 Springer

Victor Bachinsky
Bukovinian State Medical University
Chernivtsi, Ukraine

Oleh Ya Vanchulyak
Bukovinian State Medical University
Chernivtsi, Ukraine

Alexander G. Ushenko
Chernivtsi National University
Chernivtsi, Ukraine

Yurii A. Ushenko
Chernivtsi National University
Chernivtsi, Ukraine

Alexander V. Dubolazov
Chernivtsi National University
Chernivtsi, Ukraine

Alexander Bykov
Opto-Electronics and Measurement
Techniques
University of Oulu
Oulu, Finland

Benjamin Hogan
Opto-Electronics and Measurement
Techniques
University of Oulu
Oulu, Finland

Igor Meglinski
College of Engineering and Physical
Sciences
Aston University
Birmingham, UK

ISSN 2191-530X   ISSN 2191-5318 (electronic)
SpringerBriefs in Applied Sciences and Technology
ISBN 978-981-16-1449-1   ISBN 978-981-16-1450-7 (eBook)
https://doi.org/10.1007/978-981-16-1450-7

This Springer imprint is published by the registered company Springer Nature Singapore Pte Ltd.
The registered company address is: 152 Beach Road, #21-01/04 Gateway East, Singapore 189721, Singapore

Contents

Legend List

Ac	Balanced accuracy
ACI	Acute coronary insufficiency
AI	Acute ischemia
BL	Biological layers
BT	Biological tissues
CCHD	Chronic coronary heart disease
FN	False negative case
FP	False positive case
PCM	Polarisation-correlation methods
Se	Sensitivity
Sp	Specificity
TN	True negative case
TP	True positive case
λ	Wavelength

Chapter 1
Materials and Research Methods

1.1 Characterisation of Laser Polarimetric Methods

From a physical point of view, myocardium sections are optically anisotropic biological layers [1–15]. For laser radiation passing through the volume of such samples, a single averaged interaction of the laser radiation with the spatially-distributed, optically-anisotropic, network of fibrils, fibers, and other structures that form the layer is realised. As a result of this interaction, a modulation of the azimuth and polarisation ellipticity coordinates occurs in the image plane of the native myocardium section. Thus, a polarisation-inhomogeneous image of the layer is formed [15–18]. Example images of the control group, obtained on this basis using a Stokes polarimeter [13–18], are shown in Figs. 1.1, 1.2, 1.3 and 1.4 for different relative orientations of the transmission planes of the polariser and analyser.

A comparative, qualitative analysis of the polarisation structure of microscopic images of histological sections of the myocardium reveals complex manifestations of the optical anisotropy of the fibrillar myosin networks. Direct, multi-parameter studies of images of myocardium sections provide the ability to directly compare the data obtained with the results of traditional histochemical research. A systematic study of the data allows one to obtain multidimensional, objective information about changes in the optical-anisotropic structure of myocardium sections. Understanding these changes in sufficient detail can be used to accurately diagnose acute coronary insufficiency (ACI).

© The Author(s), under exclusive license to Springer Nature Singapore Pte Ltd. 2021
V. Bachinsky et al., *Multi-parameter Mueller Matrix Microscopy for the Expert Assessment of Acute Myocardium Ischemia*, SpringerBriefs in Applied Sciences and Technology, https://doi.org/10.1007/978-981-16-1450-7_1

Fig. 1.1 Microscopic image of a native myocardium section from the control group, for a relative orientation of the transmission planes of the polariser and analyser of $\Theta = 0°$

Fig. 1.2 Microscopic image of a native myocardium section from the control group, for a relative orientation of the transmission planes of the polariser and analyser of $\Theta = 45°$

1.2 Multidimensional Polarisation Microscopy of Myocardium Sections

There are a variety of possible methods for polarisation-phase microscopy of myocardium images, including:

1. Polarimetry within the set of pixels of the photosensitive area of a digital camera, with coordinate distributions of the azimuth $\alpha(p \times k)$ and ellipticity $\beta(p \times k)$ of polarisation, by polarised filtering of microscopic images of myocardium

Fig. 1.3 Microscopic image of a native myocardium section from the control group, for a relative orientation of the transmission planes of the polariser and analyser of $\Theta = 90°$

Fig. 1.4 Microscopic image of a native myocardium section from the control group, for a relative orientation of the transmission planes of the polariser and analyser of $\Theta = 135°$

sections. Following generally accepted terminology [16–18], such distributions will be called polarisation maps herein.

2. Stokes-parametric two-dimensional polarimetry of digital images of myocardium sections, based on measuring the coordinate distributions of the parameters of the Stokes vector $V_i(p \times k)$ of a microscopic image of the sample. The resulting two-dimensional distributions $V_i(p \times k)$ will be called Stokes-parametric images of the native myocardium section [1–4, 19] henceforth.

3. Mueller matrix mapping of myocardium sections, with subsequent calculation of a series of coordinate distributions of the elements of the Mueller matrix (Mueller matrix images $m_{ik}(p \times k)$), which exhaustively characterises the spatial structure of myosin networks and the degree of their crystallisation [20–27].

1.3 Methods for an Objective Analysis of Two-Dimensional Distributions of Multidimensional Polarisation Microscopy Data for Histological Sections of the Myocardium

We now consider a set of complementary methods for the objective analysis of the two-dimensional data distributions r given by multidimensional polarisation microscopy of myocardium sections. These methods are:

- statistical analysis [16–18, 28, 29];
- correlation analysis [16–18, 30];
- fractal analysis [16–18, 24, 31–33];
- wavelet analysis [16–18, 33, 34];
- and parametric analysis [35–37].

Such objective analysis methods form the basis of numerous studies of biological tissues with various structures and subject to various physiological and pathological conditions. Therefore, in this part of the work we provide only a short list and description of such methods that will be used to develop objective criteria for the forensic verification of ACI.

1.3.1 Statistical Analysis

For an objective assessment of the coordinate distributions of random variables r characterising microscopic images and morphological properties of myocardium sections, their histograms $N(r)$ are determined. From this basis, the complete set of first to fourth order statistical moments $M_{i=1;2;3;4}$ that characterise the features of the dependencies $N(r)$ is calculated.

The statistical moment of the 1st order, M_1, is the average (mean) value among the entire set (ensemble) of random values of the coordinate distributions r. The second-order statistical moment, M_2, or dispersion, characterising the distribution of a random variable r is a measure of the spread of a given quantity or its deviation from the mathematical expectation value. The third-order statistical moment, M_3, or skewness, characterises the deviation of the magnitude of the random variable r from a normal distribution. It can be calculated according to the relations given in [28]. The fourth-order statistical moment, M_4, or kurtosis, characterises the severity of the "peak" of the distribution of the random variable r of the image of the native myocardium section.

An example of the application of statistical analysis to the two-dimensional distribution of the azimuths and the ellipticity of polarisation of the microscopic image of the native myocardium section is illustrated in Figs. 1.5 and 1.6, as well as being summarised in Table 1.1.

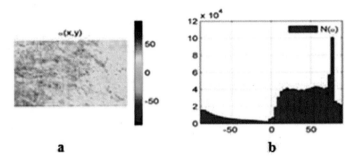

a   b

Fig. 1.5 The azimuthal dependence of the polarisation structure of the microscopic image of a section of the myocardium of the control group: **a** polarisation map of the azimuths; **b** a histogram of the distribution of random values

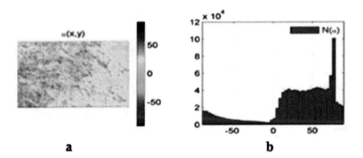

a   b

Fig. 1.6 The elliptical dependence of the polarisation structure of the microscopic image of a section of the myocardium of the control group: **a** polarisation map of the ellipticity; **b** a histogram of the distribution of random values

Table 1.1 Statistical moments M_i that characterise the distribution of azimuth and elliptic polarisation of the microscopic image of a native myocardium section from the control group

M_i	Azimuth	Ellipticity
Average (M_1)	0.68	0.08
Dispersion (M_2)	0.21	0.14
Skewness (M_3)	1.89	0.26
Kurtosis (M_4)	0.77	0.43

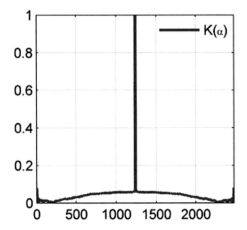

Fig. 1.7 Autocorrelation function $K(\alpha)$ of the distribution of random values of the microscopic image of a native myocardium section from the control group

1.3.2 Correlation Analysis

In order to assess the features of the coordinate structure of two-dimensional parameter distributions r, a correlation analysis of polarisation maps [38–41], Mueller matrix images [27, 42–50], or autofluorescence images [51, 52] of histological sections of the myocardium is used. This method is based on the calculation of autocorrelation functions $K(r)$ by coordinate displacement along the rows of pixels of a digital camera with coordinate distributions $r(p \times k)$ according to a known ratio. In order to quantify the autocorrelation functions $K(r)$, one can use the calculation of statistical moments of the second and fourth orders (hereinafter, the correlation moments K_2 and K_4) [30]. An example of the application of the autocorrelation analysis of the two-dimensional distribution of the azimuths and ellipticity of polarisation of the microscopic image of a native myocardium section from the control group is illustrated in Figs. 1.7 and 1.8, as well as Table 1.2.

1.3.3 Fractal Analysis

Fractal analysis provides the opportunity to obtain new information on the scale-self-similar construction of coordinate distributions of the objective parameters r. Quantitatively, this procedure is carried out according to the classical scheme; that is, by calculating the logarithmic dependences of the power spectra $\log L(r) - \log\left(1/d\right)$ of the distributions of the set of parameters r. Here, the spatial frequencies $1/d$ are determined by the geometric dimensions (d) of the distributions of polarisation states, phases, elements of the Mueller matrix, and the intensity of the autofluorescence.

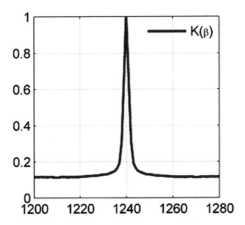

Fig. 1.8 Autocorrelation function $K(\beta)$ of the distribution of random values of the microscopic image of a native myocardium section from the control group

Table 1.2 Correlation moments M_i that characterise the autocorrelation functions of the azimuth and polarisation ellipticities of the microscopic image of a native myocardium section from the control group

K_i	Azimuth	Ellipticity
Dispersion (K_2)	0.023	0.11
Kurtosis (K_4)	1.95	0.74

In order to determine the degree of self-similarity of the coordinate distributions $r(p \times k)$, the calculated logarithmic dependences $\log L(r) - \log\left(\frac{1}{d}\right)$ are approximated by the least squares method for curves $A(u)$. The classification of the coordinate distributions r is then carried out [24, 31–33]:

- coordinate distributions r are fractal, in the case that a tilt angle u of the approximating function $A(u)$ is constant;
- coordinate distributions r are multifractal in the case of several constant tilt angles of the approximating function $A(u)$;
- coordinate distributions r are random in the case that there are no stable tilt angles of the approximating function $A(u)$ for the entire range of dimensional changes d.

An example of the application of the fractal analysis of the two-dimensional distribution of azimuths and the ellipticity of polarisation of the microscopic image of the native myocardium section is illustrated in Figs. 1.9 and 1.10.

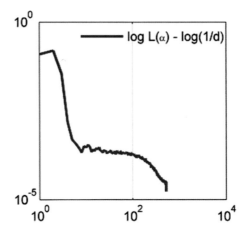

Fig. 1.9 Dependences $\log L(\alpha) - \log(d^{-1})$ of the distribution of random values of the microscopic image of a native myocardium section from the control group

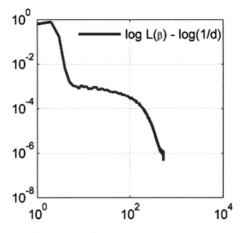

Fig. 1.10 Dependences $\log L(\beta) - \log(d^{-1})$ of the distribution of random values of the microscopic image of a native myocardium section from the control group

1.3.4 Wavelet Analysis

Among the main analytical methods for local estimation of coordinate distributions is wavelet analysis [16–18, 34, 53]. This method is based on the use of a special mathematical function, which is called a wavelet-function, as an analytical probe. Such an analytic function has a finite basis in both coordinate and frequency space. Using the wavelet-function, the distribution of the values of the calculated parameter q that characterises the image structure of the native myocardium section can be

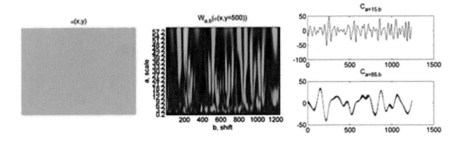

Fig. 1.11 Distribution of the wavelet-coefficients of the polarisation map of the azimuth of polarisation of the microscopic image of a native myocardium section from the control group. Here, the left panel gives the azimuthal polarisation map; the middle panel gives $W_{a,b}$ α-distribution of random values; an the right panel shows sections of wavelet-coefficients on the scales $C_{a=15,b}$ and $C_{a55,b}$

expanded into a mathematical series. This expanded series is a convolution (correlation) of the displacement parameters (b), scaling (a), and certain coefficients (wavelet coefficients). Wavelet-transforms effectively reveal both low-frequency and high-frequency characteristics of distributions r at different coordinate scales (giving a so-called "mathematical microscope"). For this reason, wavelet analysis allows us to study the multiscale structure of polarisation, phase maps, and Mueller matrix images, which are interconnected with the morphological structure of myosin myocardium networks. If we continue the analogy with the mathematical "microscope", then the displacement parameter b fixes the focal point of the microscope, and the scale factor a corresponds to the magnification.

The result of a wavelet-transform of the one-dimensional dependence of the parameter r is a two-dimensional array of amplitudes of the wavelet coefficients $W(a, b)$. The distribution of these values in space (a, b) = (spatial scale, spatial coordinate—localisation) gives information about the evolution of the relative contribution of components across different scales to the coordinate distribution. $W(a, b)$ is called the spectrum of wavelet-transform coefficients, or (frequency) scale-spatial spectrum. The spectrum is a surface in three-dimensional space.

An example of the use of wavelet-analysis of the analysis of the two-dimensional distribution of azimuths and ellipticity of polarisation of the image of the native myocardium section is illustrated in Figs. 1.11 and 1.12.

1.3.5 Parametric Analysis

The methodology for determining the dependences of the number of characteristic distribution points is as follows [16–18]:

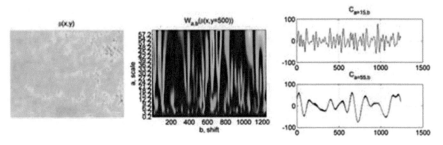

Fig. 1.12 Distribution of the wavelet-coefficients of the polarisation ellipticity map of the microscopic image of a native myocardium section from the control group. Here, the left panel gives the polarisation ellipticity map; the middle panel gives $W_{a,b}$ β-distribution of random values; and the right panel shows sections of wavelet-coefficients on the scales $C_{a=15,b}$ and $C_{a55,b}$

- discretisation of the two-dimensional distribution r of the microscopic image or map of the native myocardium section in the x direction with a step of 1 pixel;
- counting the number $N^{(1)}$ of points $r(x, y) = r_0$ within the plane $\left(1_{pix} \times p_{pix}\right)$;
- building a coordinate dependence of the number of characteristic values $N(x) = \left(N^{(1)}, N^{(2)}, \ldots, N^{(m)}\right)$.

Exemplar dependences $N(x)$ for the polarisation maps of azimuth and ellipticity of the image of a native myocardium section from the control group are illustrated in Figs. 1.13 and 1.14.

Objectively, the differences between the dependences $N(x)$ are estimated by calculating the 1st to 4th order statistical moments distributions of the number of characteristic values. Example data is given in Table 1.3.

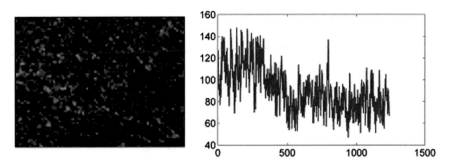

Fig. 1.13 Dependences $N(x)$ of the distribution of the polarisation azimuth of the image of a native myocardium section from the control group. Here, the left panel gives the polarisation azimuth map; the right panel gives the dependence of the number of characteristic values (0) $N(x) = \left(N^{(1)}, N^{(2)}, \ldots, N^{(m)}\right)$ of the coordinate distributions of the polarisation azimuth

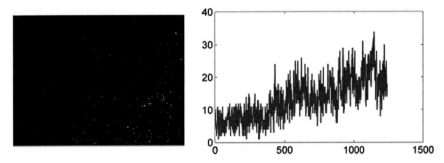

Fig. 1.14 Dependences $N(x)$ of the distribution of the polarisation ellipticity of the image of a native myocardium section from the control group. Here, the left panel gives the polarisation ellipticity map; the right panel shows the dependence of the number of characteristic values (0) $N(x) = \left(N^{(1)}, N^{(2)}, \ldots, N^{(m)}\right)$ of the coordinate distributions of the polarisation ellipticity

Table 1.3 The set of statistical moments of the 1st–4th order of the dependences $N(x)$ of the azimuth and ellipticity polarisation of the image of a native myocardium section from the control group

M_i	Azimuth	Ellipticity
Average (M_1)	0.23	0.07
Dispersion (M_2)	0.19	0.11
Skewness (M_3)	0.68	0.87
Kurtosis (M_4)	1.61	1.19

1.4 The Technique of Polarising Microscopy of Myocardium Sections

Polarisation microscopy can be performed in a standard arrangement of a Stokes-polarimeter (Fig. 1.15) [54, 55].

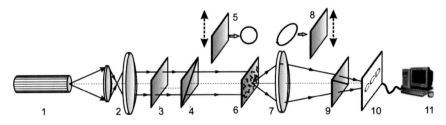

Fig. 1.15 Optical scheme of a Stokes-polarimeter. Explanation in the text of the work

Here 1–He–Ne laser; 2—collimator; 3—stationary quarter-wave plate; 4—polariser; 5—mechanically movable quarter-wave plate; 6—native section; 7—polarising micro lens; 8—mechanically movable quarter-wave plate; 9—analyser; 10—CCD camera; 11—personal computer.

The technique has been applied to the study of myocardium samples in [24]. Myocardium samples (at 6 in Fig. 1.15) were irradiated with a parallel $(\otimes = 2 \times 10^3 \mu m)$, low-intensity (P = 5.0 mW), He–Ne laser beam ($\lambda = 0.6328$ μm). The polarising irradiator consisted of two quarter-wave plates (Achromatic True Zero-Order Waveplate) (Fig. 1.15, 3, 5) and a polariser (B + W Kaesemann XS-Pro Polariser MRC Nano) (Fig. 1.15, 4). The myocardium section was sequentially probed with a laser beam with the following types of polarisation: linear with azimuths of 0°, 90°, 45° and right circular ($\otimes$). Polarisation images of a myocardium sample were projected into the plane of the photosensitive area ($m \times n = 1280 \times 960$ pixels) of the CCD camera (Fig. 2.17, 10) (The Imaging Source DMK 41AU02.AS, monochrome 1/2 "CCD, Sony ICX205AL (progressive scan)); resolution—1280 × 960; size of the photosensitive area—7600 × 6200 μm; sensitivity—0.05 lx; dynamic range—8 bit) using a polarising micro-lens 7 (Nikon CFI Achromat P, focal length—30 mm, numerical aperture—0.1, magnification—4x). Image analysis of the myocardium samples was carried out using a polariser (Fig. 1.15, 9) and a quarter-wave plate (Fig. 1.15, 8). The coordinate distributions of the parameters of the Stokes' vector $V_i(p \times k)$, azimuth $\alpha(p \times k)$, ellipticity $\beta(p \times k)$, and elements of the Mueller matrix $m_{ik}(p \times k)$ characterising the microscopic images of the myocardium sections were calculated according to well-known algorithms, the explicit forms of which are given in [56–59]. Figure 1.16 shows an example of a series of two-dimensional distributions of the parameters of the Stokes' vector of a microscopic image of a native myocardium section from the control group.

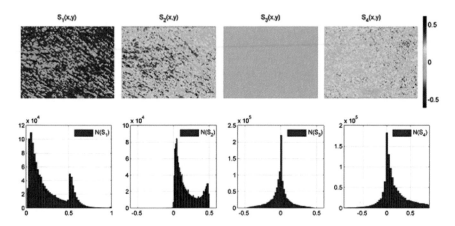

Fig. 1.16 Two-dimensional distribution of the values of the parameters of the Stokes' vector of an image of a native myocardium section from the control group

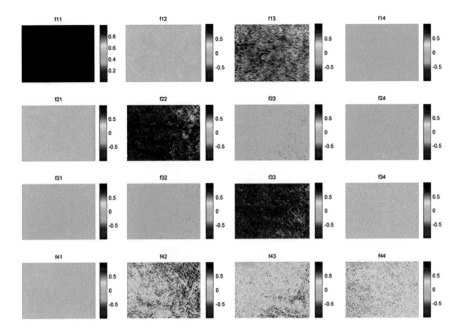

Fig. 1.17 Two-dimensional distributions of the values of the elements of the Muller matrix of a native myocardium section from the control group

In Fig. 1.17 an example of a series of two-dimensional distributions of the values of the elements of the Muller matrix of a native myocardium section from the control group is given.

Having determined a set of Stokes' parameters V_i for each pixel of the photosensitive area, one can obtain the azimuth α^* and ellipticity β^* of polarisation at the corresponding image points of the native myocardium section using known relationships [16–18].

1.5   Measurement Examples

Here we show the implementation of the investigated method for measuring two-dimensional distributions of the parameters of the Stokes' vector image of a native myocardium section using the control group as an example.

In general, for each pixel of the photosensitive area of a Stokes-polarimeter digital camera, the parameters of the Stokes' vector V of the image of the sample were determined by six measurements of the intensity following polarisation filtering. The following procedure was followed:

1.   Oriented the transmission plane of the polariser—analyser at an angle $\Theta = 0°$ and measured the intensity distribution $I_0(p \times k)$ of the microscopic image (Fig. 1.18).

2.   The polariser was rotated by an angle $\Theta = 90°$ and the coordinate distribution of the intensity $I_{90}(p \times k)$ was measured (Fig. 1.19).

Fig. 1.18 Polarising microscopic image $I_0(p \times k)$ of a native myocardium section from the control group

Fig. 1.19 Polarising microscopic image $I_{90}(p \times k)$ of a native myocardium section from the control group

Based on the definition of the Stokes' vector V [19], its first V_1 (Fig. 1.20) and second V_2 parameters (Fig. 1.21) were found:

$$V_1 = I_0 + I_{90}; \tag{1.1}$$

$$V_2 = I_0 - I_{90}. \tag{1.2}$$

3. The polariser was rotated to an angle $\Theta = 45°$ and the coordinate distribution of the intensity $I_{45}(p \times k)$ was measured (Fig. 1.22).

Fig. 1.20 Two-dimensional distribution of the first parameter of the Stokes' vector $V_1(p \times k)$ of a image of a native myocardium section from the control group

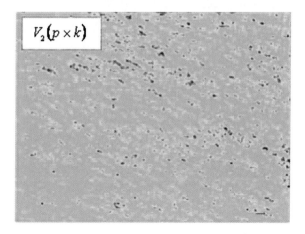

Fig. 1.21 Two-dimensional distribution of the second parameter of the Stokes' vector $V_2(p \times k)$ of a image of a native myocardium section from the control group

Fig. 1.22 Polarising microscopic image $I_{45}(p \times k)$ of a native myocardium section from the control group

4.   The polariser was rotated to an angle $\Theta = 135°$ and the coordinate distribution of the intensity $I_{135}(p \times k)$ was measured (Fig. 1.23).

5.   The third parameter V_3 of the Stokes' vector can then be calculated (Fig. 1.24).

$$V_3 = I_{45} - I_{135}. \qquad (1.3)$$

Fig. 1.23 Polarising microscopic image $I_{135}(p \times k)$ of a native myocardium section from the control group

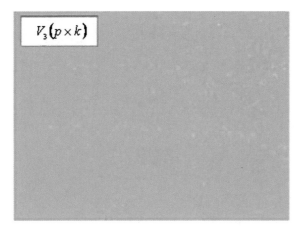

Fig. 1.24 Two-dimensional distribution of the 3rd parameter of the Stokes' vector $V_3(p \times k)$ of a image of a native myocardium section from the control group

6. To measure the fourth parameter of the Stokes' vector V_4, a quarter-wave phase plate was installed. The axis of maximum speed was oriented to an angle $0°$. The transmission plane of the analyser was oriented at an angle $\Theta = 45°$. The coordinate distribution of the intensity of the right circularly polarised radiation was measured. $I_\otimes(p \times k)$ (Fig. 1.25).

7. The transmission plane of the polariser-analyser was set at an angle $\Theta = 135^{circ}$ relative to the orientation of the axis of the maximum velocity of the phase of the quarter-wave plate. We measured the intensity distribution of left-circularly polarised radiation $I_\oplus(m \times n)$ in a microscopic image (Fig. 1.26).

Fig. 1.25 Polarising microscopic image $I_\otimes(p \times k)$ of a native myocardium section from the control group

Fig. 1.26 Polarisation image $I_\oplus(p \times k)$ of a native myocardium section from the control group

8. The coordinate distribution of the fourth parameter was calculated (Fig. 1.27)

$$V_4 = I_\otimes - I_\oplus. \tag{1.4}$$

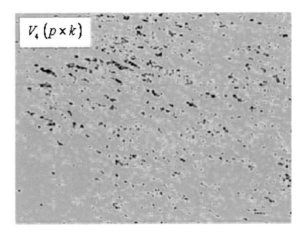

Fig. 1.27 Two-dimensional distribution of the 4th parameter of the Stokes' vector $V_4(p \times k)$ of an image of a native myocardium section from the control group

References

1. C.R. Polarimetry, *Handbook of Optics: Vol I— Geometrical and Physical Optics, Polarised Light, Components and Instruments*, ed. by M. Bass (McGraw-Hill Professional, New York, 2010), pp. 22.1–22.37.
2. N. Ghosh, M. Wood, A. Vitkin, Polarised light assessment of complex turbid media such as biological tissues via Mueller matrix decomposition, in *Handbook of Photonics for Biomedical Science* ed. by V. Tuchin (CRC Press, Taylor & Francis Group, London, 2010), pp. 253–282
3. S. Jacques, Polarised light imaging of biological tissues, in *Handbook of Biomedical Optics*. ed. by D. Boas, C. Pitris, N. Ramanujam (CRC Press, Boca Raton, London, New York, 2011), pp. 649–669
4. N. Ghosh, Tissue polarimetry: concepts, challenges, applications, and outlook. J. Biomed. Opt. **16**(11), 110801 (2011)
5. M. Swami, H. Patel, P. Gupta, Conversion of 3×3 Mueller matrix to 4×4 Mueller matrix for non-depolarising samples. Opt. Commun. **286**, 18–22 (2013)
6. D. Layde, N. Ghosh, A. Vitkin, Quantitative polarimetry for tissue characterisation and diagnosis, in *Advanced Biophotonics: Tissue Optical Sectioning*, ed. by R. Wang, V. Tuchin (CRC Press, Taylor & Francis Group, Boca Raton, London, New York, 2013), pp. 73–108.
7. T. Vo-Dinh, Biomedical Photonics Handbook: 3 volume set, 2nd ed. (CRC Press, Boca Raton, 2014)
8. A. Vitkin, N. Ghosh, A. Martino, Tissue polarimetry, in *Photonics: Scientific Foundations, Technology and Applications*, 4th edn., ed. by D. Andrews (Wiley., Hoboken, New Jersey, 2015), pp. 239–321
9. V. Tuchin, *Tissue Optics: Light Scattering Methods and Instruments for Medical Diagnosis*, 2nd edn. (SPIE Press, Bellingham, Washington, USA, 2007)
10. W. Bickel, W. Bailey, Stokes vectors, Mueller matrices, and polarised scattered light. Am. J. Phys. **53**(5), 468–478 (1985)
11. A. Doronin, C. Macdonald, I. Meglinski, Propagation of coherent polarised light in turbid highly scattering medium. J. Biomed. Opt. **19**(2), 025005 (2014)
12. A. Doronin, A. Radosevich, V. Backman, I. Meglinski, Two electric field monte carlo models of coherent backscattering of polarised light. J. Opt. Soc. Am. A **31**(11), 2394 (2014)
13. P. Arun Gopinathan, G. Kokila, M. Jyothi, C. Ananjan, L. Pradeep, S.H. Nazir, Study of collagen birefringence in different grades of oral squamous cell carcinoma using picrosirius red and polarised light microscopy. Scientifica **2015**, 802980 (2015)
14. L. Rich, P. Whittaker, Collagen and picrosirius red staining: a polarised light assessment of fibrillar hue and spatial distriburuon. Braz. J. Morphol. Sci. (2005)
15. S. Bancelin, A. Nazac, B.H. Ibrahim et al., Determination of collagen fiber orientation in histological slides using Mueller microscopy and validation by second harmonic generation imaging. Opt. Express **22**(19), 22561–22574 (2014)
16. A. Ushenko, V. Pishak, Laser polarimetry of biological tissue: principles and applications, in *Handbook of Coherent-Domain Optical Methods: Biomedical Diagnostics*, ed. by V. Tuchin (Environmental and Material Science, 2004), pp. 93–138
17. O. Angelsky, A. Ushenko, Y. Ushenko, V. Pishak, A. Peresunko, Statistical, correlation and topological approaches in diagnostics of the structure and physiological state of birefringent biological tissues. Handbook of Photon. Biomed. Sci. (2010), pp. 283–322
18. Y. Ushenko, T. Boychuk, V. Bachynsky, O. Mincer, Diagnostics of structure and physiological state of birefringent biological tissues: statistical, correlation and topological approaches, in *Handbook of Coherent-Domain Optical Methods*, ed. by V. Tuchin (Springer Science+Business Media; 2013)
19. V. Tuchin, L. Wang, D. Zimnjakov, *Optical Polarisation in Biomedical Applications* (Springer, New York, USA, 2006).
20. Y. Ushenko, V. Ushenko, A. Dubolazov, V. Balanetskaya, N. Zabolotna, Mueller matrix diagnostics of optical properties of polycrystalline networks of human blood plasma. Opt. Spectrosc. **112**(6), 884–892 (2012)

21. V. Ushenko, O. Dubolazov, A. Karachevtsev, Two wavelength Mueller matrix reconstruction of blood plasma films polycrystalline structure in diagnostics of breast cancer. Appl. Opt. **53**(10), B128 (2016)
22. Y. Ushenko, G. Koval, A. Ushenko, O. Dubolazov, V. Ushenko, O. Novakovskaia, Mueller matrix of laser-induced autofluorescence of polycrystalline films of dried peritoneal fluid in diagnostics of endometriosis. J. Biomed. Opt. **21**(7), 071116 (2016)
23. O.V. Angelsky, Y.Y. Tomka, A.G. Ushenko, Y.G. Ushenko, Y.A. Ushenko, Investigation of 2D Mueller matrix structure of biological tissues for pre-clinical diagnostics of their pathological states. J. Phys. D: Appl. Phys, **38**, 4227–4235 (2005)
24. Yu.A. Ushenko, Fractal structure of Mueller matrices images of biotissues. Proc. SPIE **5772**, 131–138 (2004)
25. V.O. Ushenko, O. Vanchuliak, M.Yu. Sakhnovskiy, O.V. Dubolazov, P. Grygoryshyn, I.V. Soltys, O.V. Olar, System of Mueller matrix polarisation correlometry of biological polycrystalline layers. Proc. SPIE Int. Soc. Opt. Eng. **10352**, 103520U (2017)
26. O.V. Dubolazov, V.O. Ushenko, L. Trifoniuk, Y.O. Ushenko, V.G. Zhytaryuk, O.G. Prydiy, M. Grytsyuk, L. Kushnerik, I. Meglinskiy, Methods and means of 3D diffuse Mueller matrix tomography of depolarising optically anisotropic biological layers. Proc SPIE Int. Soc. Opt. Eng. **10396**, 103962P (2017)
27. A.G. Ushenko, A.V. Dubolazov, V.A. Ushenko, Yu.A. Ushenko, L.Ya. Kushnerik, O.V. Olar, N.V. Pashkovskaya, Yu.F. Marchuk, Mueller matrix differentiation of fibrillar networks of biological tissues with different phase and amplitude anisotropy. Proc SPIE Int. Soc. Opt. Eng. **9971**, 99712K (2016)
28. A. Ushenko, A. Dubolazov, V. Ushenko, O. Novakovskaya, Statistical analysis of polarisation-inhomogeneous Fourier spectra of laser radiation scattered by human skin in the tasks of differentiation of benign and malignant formations. J. Biomed. Opt. **21**(7), 071110 (2016)
29. Yu.O. Ushenko, O.V. Dubolazov, V.O. Ushenko, V.G. Zhytaryuk, O.G. Prydiy, N. Pavlyukovich, O. Pavlyukovich, Statistical analysis of polarisation interference images of biological fluids polycrystalline films in the tasks of optical anisotropy weak changes differentiation . Proc. SPIE Int. Soc. Opt. Eng. **10612**, 106121Q (2018)
30. O. Angelsky, A. Ushenko, Y. Ushenko, Investigation of the correlation structure of biological tissue polarisation images during the diagnostics of their oncological changes. Phys. Med. Biol. **50**(20), 4811–4822 (2005)
31. B.B. Mandelbrot, On the geometry of homogeneous turbulence, with stress on the fractal dimension of the iso-surfaces of scalars. J. Fluid Mech. **72**, 401 (1975)
32. B.B. Mandelbrot, *The Fractal Geometry of Nature* (W.H.Freeman, San Francisco, 1982), p. 460
33. E.L. Church, Fractal surface finish. Appl. Opt **27**, 1518–1526 (1988)
34. O.Y. Wanchulyak, V.T. Bachinsky, A.G. Ushenko et al., Wavelet analysis of dynamics of changes in orientation-phase structures biotissue architectonics. Proc. SPIE **5067**, 50–55 (2002)
35. N. Sivokorovskaya, V.T. Bachinskyi, O.Y. Vanchulyak, O.G. Ushenko, A.V. Dubolazov, Y.O. Ushenko, Y.Y. Tomka, L.Y. Kushnerik, Statistical analysis of polarisation images of histological cuts of parenchymatic tissues in diagnostics of volume of blood loss. IFMBE Proc. **77**, 513–517 (2020)
36. O.G. Ushenko, A.-V. Syvokorovskaya, V.T. Bachinsky, O.Y. Vanchuliak, A.V. Dubolazov, Y.O. Ushenko, Y.Y. Tomka, M.L. Kovalchuk, Laser autofluorescent microscopy of histological sections of parenchymatous biological tissues of the dead. IFMBE Proc. **77**, 507–511
37. Y. Sarkisova, V.T. Bachinskyi, M. Garazdyuk, O.Y. Vanchulyak, O.Y. Litvinenko, O.G. Ushenko, B.G. Bodnar, A.V. Dubolazov, Y.O. Ushenko, Y.Y. Tomka, I.V. Soltys, S. Foglinskiy, Differential muller-matrix microscopy of protein fractions of vitreous preparations in diagnostics of the pressure of death. IFMBE Proc. **77**, 503–506 (2020)
38. A.G. Ushenko, A.V. Dubolazov, V.A. Ushenko, Yu.A. Ushenko, M.Yu. Sakhnovskiy, O.I. Olar, Methods and means of laser polarimetry microscopy of optically anisotropic biological layers. Proc. SPIE Int. Soc. Opt. Eng. **9971**, 99712B (2016)
39. M.Y. Sakhnovskiy, A.-V. Syvokorovskaya, V. Martseniak, B.M. Bodnar, O. Tsyhykalo, A.V. Dubolazov, O.I. Olar, V.A. Ushenko, P.M. Grygoryshyn, System of biological crystals fibrillar

networks polarisation-correlation mapping . Proc. SPIE Int. Soc. Opt. Eng. **10752**, 107522G (2018)

40. M.Y. Sakhnovskiy, O.Y. Wanchuliak, B. Bodnar, I.V. Martseniak, O. Tsyhykalo, A.V. Dubolazov, V.A. Ushenko, O.I. Olar, P.M. Grygoryshyn, Polarisation-interference images of optically anisotropic biological layers. Proc. SPIE Int. Soc. Opt. Eng. **10752**, 107522F (2018)

41. Y.A. Ushenko, O.V. Olar, A.V. Dubolazov, O.B. Bodnar, B.M. Bodnar, L. Pidkamin, O. Prydiy, M.I. Sidor, D. Kvasnyuk, O. Tsyhykalo, System of differential Mueller matrix mapping of phase and amplitude anisotropy of depolarising biological tissues. Proc. SPIE Int. Soc. Opt. Eng. **10752**, 107522H (2018)

42. O.V. Dubolazov, A.G. Ushenko, Y.A. Ushenko, M.Y. Sakhnovskiy, P.M. Grygoryshyn, N. Pavlyukovich, O.V. Pavlyukovich, V.T. Bachynskiy, S.V. Pavlov, V.D. Mishalov, Z. Omiotek, O, Mamyrbaev, Laser Müller matrix diagnostics of changes in the optical anisotropy of biological tissues information technology in medical diagnostics, in *II—Proceedings of the International Scientific Internet Conference on Computer Graphics and Image Processing and 48th International Scientific and Practical Conference on Application of Lasers in Medicine and Biology, 2018* (2019), pp. 195–203

43. M. Borovkova, M. Peyvasteh, O. Dubolazov, Y. Ushenko, V. Ushenko, A. Bykov, S. Deby, J. Rehbinder, T. Novikova, I. Meglinski, Complementary analysis of Mueller matrix images of optically anisotropic highly scattering biological tissues. J. Eur. Opt. Soc. **14**(1), 20 (2018)

44. V. Ushenko, A. Sdobnov, A. Syvokorovskaya, A. Dubolazov, O. Vanchulyak, A. Ushenko, Y. Ushenko, M. Gorsky, M. Sidor, A. Bykov, I. Meglinski, 3D Mueller matrix diffusive tomography of polycrystalline blood films for cancer diagnosis. Photonics **5**(4), 54 (2018)

45. L. Trifonyuk, W. Baranowski, V. Ushenko, O. Olar, A. Dubolazov, Y. Ushenko, B. Bodnar, O. Vanchulyak, L. Kushnerik, M. Sakhnovskiy, 2D-Mueller matrix tomography of optically anisotropic polycrystalline networks of biological tissues histological sections. Opto-Electron. Rev. **26** (3), 252–259 (2018)

46. V.A. Ushenko, A.V. Dubolazov, L.Y. Pidkamin, M.Y. Sakchnovsky, A.B. Bodnar, Y.A. Ushenko, A.G. Ushenko, A. Bykov, I. Meglinski, Mapping of polycrystalline films of biological fluids utilising the Jones-matrix formalism. Laser Phys. **28**(2), 025602 (2018)

47. Y.A. Ushenko, A.V. Dubolazov, O.V. Olar, S.O. Sokolnyk, G.B. Bodnar, L. Pidkamin, O. Prydiy, M.I. Sidor, Clinical applications of the Mueller matrix reconstruction of the polycrystalline structure of multiple-scattering biological tissues . Proc. SPIE Int. Soc. Opt. Eng. **10728**, 107280P (2018)

48. L. Trifonyuk, V. Baranovsky, O.V. Dubolazov, V.O. Ushenko, O.G. Ushenko, V.G. Zhytaryuk, O.G. Prydiy, O. Vanchulyak, Jones-matrix tomography of biological tissues phase anisotropy in the diagnosis of uterus wall prolapse . Proc. SPIE Int. Soc. Opt. Eng. **10612**, 106121F (2018)

49. Y.A. Ushenko, A.V. Dubolazov, O.B. Bodnar, B.M. Bodnar, L. Pidkamin, O. Prydiy, M.I. Sidor, I.V. Martseniak, O. Tsyhykalo, Holographic reconstruction of optical anisotropy of blood films and diagnostics of prostate cancer. Proc. SPIE Int. Soc. Opt. Eng. **10977**, 109773S (2018)

50. Y.A. Ushenko, O. Bakun, I.V. Martseniak, O. Tsyhykalo, A.V. Dubolazov, L.Y. Pidkamin, O.G. Prydiy, I.V. Soltys, M.P. Gorsky, Polarizarion reconstruction of polycrystalline structure of biological liquid films . Proc. SPIE Int. Soc. Opt. Eng. **10977**, 109773R (2018)

51. V.A. Ushenko, A.Y. Sdobnov, W.D. Mishalov, A.V. Dubolazov, O.V. Olar, V.T. Bachinskyi, A.G. Ushenko, Y.A. Ushenko, O.Y. Wanchuliak, I. Meglinski, Biomedical applications of Jones-matrix tomography to polycrystalline films of biological fluids. J. Innovat Opt. Health Sci. **12** (6), 1950017 (2019)

52. M. Borovkova, L. Trifonyuk, V. Ushenko, O. Dubolazov, O. Vanchulyak, G. Bodnar, Y. Ushenko, O. Olar, O. Ushenko, M. Sakhnovskiy, A. Bykov, I. Meglinski, Mueller matrix-based polarisation imaging and quantitative assessment of optically anisotropic polycrystalline networks PLoS ONE, **14**(5), e0214494 (2019)

53. A.G. Ushenko, D.N. Burkovets, Wavelet-analysis of two-dimensional birefringence images of architectonics in biotissues for the diagnostics of pathological changes. J. Biomed. Opt **9**(4), 1023–1028 (2004)

54. O. Angelsky, Y. Tomka, A. Ushenko, Y. Ushenko, S. Yermolenko, 2-D tomography of biotissue images in pre-clinic diagnostics of their pre-cancer states. Proc. SPIE **5972**, 158–162 (2005)
55. V. Prysyazhnyuk, Y. Ushenko, A. Dubolazov, A. Ushenko, V. Ushenko, Polarisation-dependent laser autofluorescence of the polycrystalline networks of blood plasma films in the task of liver pathology differentiation. Appl. Opt. **55**(12), B126 (2016)
56. Y.A. Ushenko, G.D. Koval, A.G. Ushenko, O.V. Dubolazov, V.A. Ushenko, O.Y. Novakovskaia, Mueller matrix of laser-induced autofluorescence of polycrystalline films of dried peritoneal fluid in diagnostics of endometriosis. J. Biomed. Opt. **21** (7), 071116
57. V.P. Prysyazhnyuk, Yu.A. Ushenko, A.V. Dubolazov, A.G. Ushenko, V.A. Ushenko, Polarisation-dependent laser autofluorescence of the polycrystalline networks of blood plasma films in the task of liver pathology differentiation. Applied Optics, **55** (12), B126-B132 (2016)
58. A.G. Ushenko, O.V. Dubolazov, V.A. Ushenko, OYu. Novakovskaya, O.V. Olar, Fourier polarimetry of human skin in the tasks of differentiation of benign and malignant formations. Appl. Opt. **55**(12), B56–B60 (2016)
59. Yu.A. Ushenko, V.T. Bachynsky, O.Ya. Vanchulyak, , A.V. Dubolazov, M.S. Garazdyuk, V.A. Ushenko, Jones-matrix mapping of complex degree of mutual anisotropy of birefringent protein networks during the differentiation of myocardium necrotic changes. Appl. Opt. **55**(12), B113–B119 (2016)

Chapter 2
Scale-Selective Multidimentional Polarisation Microscopy in the Post-mortem Diagnosis of Acute Myocardium Ischemia

2.1 Model Consideration of Optical Myocardium Anisotropy

In a series of works [1–7], the following approach was proposed and successfully tested for problems in forensic medicine. The approach allows one to model the optical manifestations of the morphological structure of biological tissues (BT). It is based upon the following assumptions:

1. The whole variety of human BT can be generally represented as one of four main types—connective, muscle, epithelial, and nervous tissue.
2. The morphological structure of any type of BT is considered as a two-component amorphous-crystalline structure.
3. The crystalline component or extracellular matrix is an architectonic network consisting of protein (collagen, myosin, elastin, etc.) fibrils.

From a morphological point of view, the heart is a hollow muscle organ that is located in the chest cavity, within the mediastinum. Anatomically, we can distinguish the apex of the heart and the base of the heart, the sternocostal surface, the anterior, diaphragmatic, lower surface and the right/left pulmonary surface. An external examination of the heart clearly outlines the coronary sulcus, which is a projection of the border between the ventricles and the atria of the heart. From the coronary sulcus, on the front and bottom surfaces pass the anterior interventricular sulcus and posterior interventricular sulcus, which are the projection of the border between the right and left ventricles, passed along the left and right interventricular arteries and their accompanying veins. On the inner surface of the right ventricle (ventriculus dexter) there are fleshy septa that form the anterior papillary muscle, the posterior papillary muscle and the septal papillary muscle. The left ventricle is the largest chamber of the heart and forms most of its diaphragm surface. Fleshy septa form the anterior and posterior mastoid muscles [8–10].

V. Bachinsky et al., *Multi-parameter Mueller Matrix Microscopy for the Expert Assessment of Acute Myocardium Ischemia*, SpringerBriefs in Applied Sciences and Technology, https://doi.org/10.1007/978-981-16-1450-7_2

The wall of the heart consists of the endocardium, myocardium and epicardium. The muscle fibers of the atria and ventricles begin from the fibrous tissue, which is part of the soft skeleton of the heart. The atrial myocardium consists of a surface layer with ring fibers, which is common to both atria and a deep layer with longitudinal bundles, which is separate for each of the atria. The ventricular myocardium consists of: the outer layer, which starts from the fibrous rings, has a longitudinal direction of the fibers, and continues down to the apex of the heart, where it forms a curl of the heart, passing into the deep layer on the opposite side; the middle layer with a circular arrangement of fibers, which is separate for each ventricle; the deep layer with a longitudinal direction of the fibers [10].

The myocardium is a combination of highly specialised muscle cells—cardiomyocytes and interstitiums. The length of cardiomyocytes from different parts of the heart ranges from 50 to 120 microns, and they have widths of 10–20 microns. Using "end-to-end" contacts, cardiomyocytes are combined into muscle fibers with well-defined transverse banding. Skew transverse bridges (nexus) between their lateral surfaces integrate the myocardium into the "functional syncytium" [11–19].

At the micro- and macro-scopic levels, there is a clear tendency toward the unification of muscle fibers into bundles (fascicules) of various powers. The gentle layers of loose connective tissue that limit them are the distribution of intramural vessels and lymphatic network elements. However, the allocation of the myocardium bundle structure is conditional due to the wide exchange of fascicles by the muscle fibers and microvessels that accompany them [19]. The structural consolidation of muscle fibers and the patency of the capillaries that feed them in all phases of the cardiac cycle is ensured by the collagen framework, which has three levels of organisation. The collagen network covers entire groups of muscle fibers and intramural vessels, surrounds microvessels, intermyocytic and myocytic-capillary compounds, forming a complex three-dimensional system. The endomysium is a delicate discontinuous case of bundles of collagen filaments 120–150 nm thick, firmly connected to cardiomyocytes at the level of their Z-lines. This connection is carried out by short strands that are formed around the cuff cells in contact with specialised adhesion molecules with cytoskeletal elements. The plexus perimisium is joined by transverse collagen fibers and coarser longitudinally oriented fibrous structures. The elastic fibers in the myocardium are single, their number increases at an older age [9, 10].

The components of the interstitium are correlated as follows: 55% are vascular structures, 5% are elements of the nervous system, connective tissue cells, its fibers and proteoglycans account for 7%, 4% and 23% of the volume, respectively, and the remaining 6% are optically empty zones. The loose connective tissue of the interstitium, which structurally integrates the myocardium, is a multifunctional system of interconnected elements [10].

The intercellular substance provides an environment for the normal functioning of the working components of the myocardium. The intercellular substance includes a highly hydrated gel, the transport-trophic properties of which are determined by proteoglycans and glycoproteins. Their large molecules consist of a protein covalently linked to glycosaminoglycans. The most important of these are hyaluronic acid, heparin, sulfated and non-sulfated chondroitin. The number of fibrous elements in the interstitium of a healthy myocardium is negligible. Their basis is collagen of types I and, in a smaller amount, III. Collagen molecules are layered ordered structures that have the properties of liquid crystals. Collagen fibers and intercellular substance are oriented, as a rule, parallel to the tensile force. This alignment ensures the maximum efficiency of their supporting-skeletal function. Connected collagen fibrils spiral around the cardiomyocytes, preventing the overstretching of muscle fibers and limiting their mutual displacement during the dynamics of the cardiac cycle. The deformations of collagen structures that arise in this case generate additional force when the heart returns to its original volume.

The different substances structured in collagen fibers, basal membranes, and intercellular matter are produced by various cells. The synthesis of glycosaminoglycans and scleroproteins is carried out in the endoplasmic reticulum and Golgi complex of fibroblasts, in the vascular wall—by smooth muscle cells. Heparin is synthesised in mast cells.

In the myocardium, cells of the fibroblastic series differ in the degree of maturity and specialisation. All of them are capable of migration in a three-dimensional collagen gel. Glycolysis is the basis of their energy metabolism. The synthesised collagen molecules have the ability to self-assemble, the order of which is determined by the arrangement of amino acids in the final sections. This process and fiber formation modeling are carried out on the surface of fibroblasts, which does not exclude "remote" fibrillation however.

Under physiological conditions, collagen catabolism, like fibrillogenesis, is carried out on a cellular basis. Desmolytic factors—cathepsin, collagenase, hyalunidases, a number of non-lysosomal enzymes are produced not only by specialised fibroblasts, but also by macrophages, mast cells, and leukocytes. Enzymatic disintegration of the intercellular substance is regulated by changes in pH, Ca^{2+} concentration, and haematological factors. Cellular resorption by fibroblasts of kurtosis or defective collagen is immuno-independent and is regulated by "collagen-sensitive" receptors located on their surface. The balance of desmolytic and desmoplastic processes is an important factor in tissue homeostasis [17, 18].

Fibroblasts, macrophages, and sometimes blood cells, present in the interstitium serve as a source of numerous cytokines that not only regulate stromal morphogenesis, but also affect the endothelium and cardiomyocytes. However, the cardiomyocytes themselves have the most powerful effect on the morphofunctional state of all tissue components that maintain a normal balance of stroma and contractile myocardium. The spectrum of biologically active substances synthesised by ventricular cardiomyocytes includes metalloproteases, collagenases (stromelin, gelatinases A and B) and a number of other humoral factors [16].

Cardiomyocytes directly contain all organelles of a general nature, but the degree of their development is different. The nucleus, usually bright, is characterised by the presence of a large amount of euchromatin. In humans, the number of binuclear myositis does not exceed 10–13%. The Golgi apparatus is represented by 3–4 cisterns and an accumulation of small bubbbles and vesicles, usually located at the poles of the nucleus. Lysosomes are predominantly located around the nucleus. Granular endoplasmic reticulum is developed slightly—sometimes there are single channels.

Mitochondria, from the number of organelles of a general nature, have reached the greatest development and occupy a significant volume of cardiomyocytes of the ventricles of the heart. A characteristic feature of mitochondria is the presence of specific structures—intermitochondral contacts. Since a cardiomyocyte belongs to excitable cells, its cytoplasmic membrane specialises in conducting (and, in the case of cardiomyocytes of the cardiac conduction system, generating) an action potential, which is reflected in the large number of ion channels [13]. Some of them (calcium) are quite specific. To transform an electrical signal into a contraction of myofibrils, the cytolemma forms a concavity—a system of T-tubules. Inside the cardiomyocyte, the Golgi apparatus specialises in sarcoplasmic reticulum forming L-tubules. The contact points of neighbouring cells in the functional fiber are perceived as clearer transverse stripes—insertable discs. In the area of insertion disks between cardiomyocytes there are interdigitations, desmosomes, and nexuses. Functional fibers are surrounded by a basement membrane in such a way that it covers only the lateral surfaces of cardiomyocytes but does not go to their base (end surfaces). Reduction is provided by actin-myosin disks with anisotropic properties [20–25].

The description of the mechanisms of optical myocardium anisotropy is based on the following model representations [26–32]:

1. Amino acids and the myosin polypeptide chains formed by them (primary protein structure) have *optical activity* [20–25, 33–42]. Such a mechanism leads to the rotation of the plane of polarisation of the laser beam. Due to the peculiarities of the coordinate distribution of polypeptide chains in the plane of the histological section of the myocardium, a multitude of rotations of the plane of polarisation at different points of the corresponding image are formed. By recording the polarisation map of the azimuth, we can obtain information about the manifestations of optically active structures of the primary structure of proteins that form the morphological structure of the myocardium.

2. Fibrillar (secondary structure) protein networks that are formed by polypeptide chains have *linear birefringence* [20–25, 33–42]. An optical manifestation of the features of the morphological structure of the fibrillar mesh is the formation of a coordinate distribution of the intensity in the image plane of the histological section of the myocardium. Thus, by registering the polarisation map of the ellipticity of polarisation, one can obtain information on the manifestations of the properties of fibrillar networks (networks) that form the secondary structure of the morphological structure of the myocardium.

3. Ischemic changes in the morphological structure of the myocardium lead to structural and biochemical transformation of the primary and secondary structure of its protein components. Optically, such processes lead to changes in the coordinate distributions of azimuth maps (biochemical processes) and ellipticity (orientational changes in fibrillar networks) of the polarisation of images of myocardium sections.

4. The main idea of establishing ACI and its differentiation lies in the possibility

of multi-parameter $\left(r \equiv \left[\begin{pmatrix} V_i \\ \alpha \\ \beta \\ m_{ik} \end{pmatrix}, \begin{pmatrix} I \\ \alpha^* \\ \beta^* \end{pmatrix} \right] \right)$ objectification of the analysis

of image sections.

5. By cross-statistical analysis of three groups of samples (control (group 1), chronic coronary heart disease (CCHD, group 2) and ACI (group 3)) using the methodology of evidence-based medicine, the diagnostic strength of the polarisation microscopy method in post-mortem diagnosis of myocardium sections is determined.

2.2 Verification of Acute Coronary Insufficiency and Differentiation from Chronic Coronary Heart Disease Using Multidimensional Polarisation Microscopy of Myocardium Sections

2.2.1 Azimuth Maps of Polarisation of Images of Sections of the Myocardium of the Studied Groups

Measurement of polarisation maps of the azimuth of polarisation of myocardium images was carried out using a Stokes polarimeter [1, 43].

By rotating the axis of transmission of the analyser 9 by an angle Θ in the range $= 0°–180°$, arrays of the minimum and maximum levels of the intensity $I_{min}(p \times k)$; $I_{max}(p \times k)$ of the image of the biological object for each individual pixel (m, n) of the CCD camera and the corresponding rotation angles $\Theta(p \times k)(I(p \times k) \equiv \min())$ are determined. Next, the coordinate distributions (polarisation maps) of the azimuths of the polarisation of the image of the biological object are calculated using the relation:

$$\alpha(p \times k) = \Theta\left(I(p \times k) \equiv \min() \frac{\pi}{2}\right). \tag{2.1}$$

Figure 2.1 illustrates a series of polarisation images of an optically anisotropic matrix (in crossed $\Delta\Theta = 90^0$ transmission planes of the polariser and analyser) of myocardium tissue samples from each group.

A comparative analysis of polarised images of the myocardium samples revealed that, regardless of the cause of death, optically anisotropic primary and secondary structures of the proteins of the fibrillar networks of the actinomyosin complex make a significant contribution to the formation of polarisation azimuths.

This fact confirms the presence of a large number of light regions that are polarised-transformed due to the influence of optical anisotropy (activity of polypeptide chains) of myosin molecules in images of sections of myocardium samples of both groups.

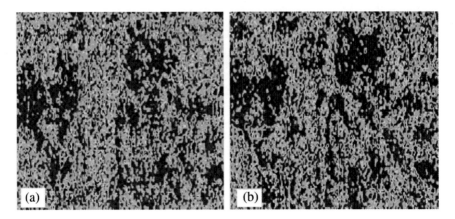

Fig. 2.1 Laser images of the polarised-visualised fibrillar mesh of the myocardium section: **a** ACI; **b** CCHD

A microscopic image of a group 2 myocardium tissue sample (Fig. 2.1a) is characterised by a fairly uniform distribution of the light areas corresponding to the directions of laying of myosin myocardium fibrils.

Morphological changes in myocardium tissue due to ACI are manifested in the formation of local clusters of anisotropic structures—in the corresponding visualised polarisation images of the fibrillar networks of its sections (Fig. 2.1b), a certain coordinate localisation of bright areas is observed.

It should be expected that such qualitatively analysed optical manifestations of the tendency to change the morphological structure of myocardium tissue due to various causes of death will more definitively be recognised in the differences in the values and ranges of changes in statistical moments of the 1st–4th orders of magnitude characterising the distribution of polarisation azimuths (relation 2.1) of the polarised images of the studied samples.

The distributions $\alpha(p \times k)$ (fragment a) and histograms (fragment b) of their values calculated for the polarisation image of a group 2 myocardium tissue sample are shown in Fig. 2.2.

An analysis of the polarisation map of azimuths (Fig. 2.2a) of the polarisation image of a section of the myocardium tissue from group 2 found that the histogram of random values α is characterised by a maximum range of variation ($0° \leq \Delta\alpha \leq 180°$) of the values of the azimuth of the polarisation of laser radiation. However, the ratio of the magnitude of the main extremum ($\alpha_0 = 90°$) and local extrema is 1–50 (Fig. 2.2b). This circumstance objectively determines a high level of values of the statistical moment of the 4th order, which characterises the distribution of polarisation azimuths of the microscopic image of myocardium tissue.

On the other hand, the histogram of the distribution of the azimuths of polarisation of the image of the group 2 myocardium tissue is somewhat asymmetric with respect to the main extremum. Statistically, this will be observed in the corresponding value

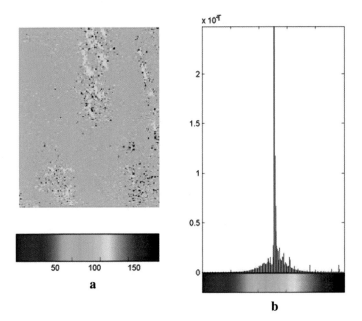

Fig. 2.2   Laser imaging of a section of myocardium tissue with ACI. **a** two-dimensional $(p \times k)$ distribution of polarisation azimuths α; **b** histograms of random polarisation azimuths

of the statistical moment of the third order, characterising the skewness of the distribution of the azimuth values of the polarisation map $\alpha(p \times k)$ of the polarisation image.

Quantitatively, the optical manifestations of the anisotropy of the networks of myosin fibrils of myocardium tissue during ACI are illustrated by a set of values of statistical moments of the 1st-4th order $M_1^{\alpha} = 0.56$; $M_2^{\alpha} = 0.16$; $M_3^{\alpha} = 0.28$; $M_4^{\alpha} = 1.19$.

Figure 2.3 shows the results of a polarisation map $\alpha(p \times k)$ of a microscopic image of a histological section of human myocardium tissue with ACI.

The information obtained about the coordinate (Fig. 2.3a) and statistical (Fig. 2.3b) distribution of rotation values of the plane of polarisation of laser radiation at points of a digital microscopic image indicates a certain decrease in the optical activity of the substance of myosin fibrils of a sample of myocardium tissue group with CCHD. As can be seen, the magnitude of the main extremum of the polarisation azimuth $\alpha_0 = 90°$ values increases compared to the tissue with ACI. In addition, the distribution of random values of the azimuth of polarisation becomes more symmetrical.

Quantitatively, changes in the biochemical and geometric structure of protein fibrils characterise the change of magnitudes in the statistical moments of the 3rd (decrease by 40%) and 4th orders (increase by 35%), which in turn characterise the coordinate distribution of the azimuth of the polarisation map of the polarimetric

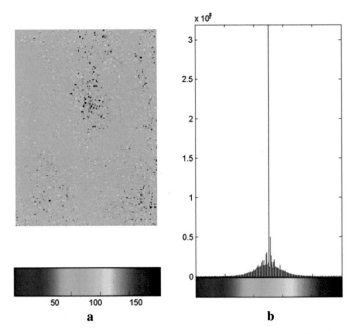

Fig. 2.3 Laser image of a section of myocardium tissue in CCHD. Here **a** two-dimensional ($p \times k$) distribution of polarisation azimuths α; **b** histograms of random polarisation azimuths

image of myocardium tissue of a person who died due to ACI,—$M_1^\alpha = 0.59$; $M_2^\alpha = 0.14$; $M_3^\alpha = 0.18$; $M_4^\alpha = 1.61$.

The correlation structure of polarisation maps $\alpha(p \times k)$ of images of sections of myocardium tissue at ACI and CCHD is illustrated by the autocorrelation functions $K^\alpha(\Delta p)$, which are shown in Fig. 2.4.

From the data obtained, it is seen that the coordinate distribution of the values of the polarisation azimuths of the laser image of the histological section of myocardium

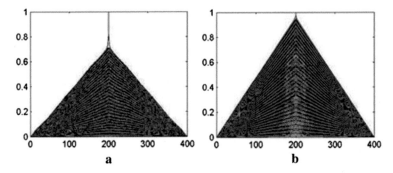

Fig. 2.4 Autocorrelation functions of the distribution of polarisation azimuths in images of sections of myocardium tissue: **a** at ACI; **b** at CCHD

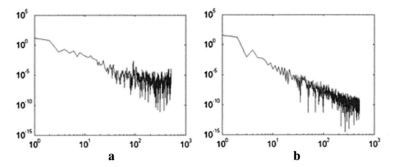

Fig. 2.5 Dependences log $L(\alpha) - \log 1/d$ of microscopic images of myocardium sections: **a** ACI; **b** CCHD

tissue at ACI is more uniform. This follows from the fact that the autocorrelation function $K^{\alpha}(\Delta p)$ calculated for the polarisation map $\alpha(p \times k)$ of the microscopic image of the histological section of myocardium tissue with ACI (Fig. 2.4a) decreases faster than the similar correlation function that was obtained for the myocardium with CCHD (Fig. 2.4b). Figure 2.5 shows the logarithmic dependences of the power spectra of the distribution of polarisation azimuths of microscopic images of sections of myocardium tissue with ACI and CCHD.

It can be seen from the obtained data that the coordinate distributions of the polarisation azimuths in the plane of microscopic images of sections of myocardium tissue at ACI and CCHD are different.

For the myocardium with ACI, the distribution $\alpha(p \times k)$ is fractal, since the dependence $\log L(\alpha) - \log 1/d$ is characterised by a constant slope $u = const$ of the approximating curve $A(u)$.

For the myocardium with CCHD, the distribution $\alpha(p \times k)$ is statistical in the small size range (2–20 µm) of myosin myocardium fibrils, since for approximating curves $A(u)$ there is no stable value of the angle of inclination $u \neq const$.

2.2.2 Maps of Polarisation Ellipticity of Microscopic Images of Myocardium Sections with Chronic Coronary Heart Disease and Acute Coronary Insufficiency

The technique for measuring the coordinate distribution of polarisation ellipticity in microscopic digital images of myocardium sections follows these steps:

- by rotating the axis of transmission of the analyser by an angle Θ in the range $= 0°$–$180°$, arrays of minimum and maximum intensity levels $I(p \times k)(p \times k)_{max_{min}}$ of the microscopic image of the myocardium section for each individual pixel (pk) of the CCD camera and the corresponding rotation angles $\Theta(p \times k)(I(p \times k) \equiv min())$ are determined;

- further, coordinate distributions (polarisation maps) of the polarisation ellipticity of the microscopic image of the myocardium section are calculated using the ratio:

$$\beta(p \times k) = arctg \frac{I(p \times k)_{min}}{I(p \times k)_{max}}. \qquad (2.2)$$

Figure 2.6 shows the coordinate distribution of polarisation ellipticity β and histograms of their values in a microscopic image of a histological section of the myocardium with ACI.

The presence of birefringence of the substance of myosin fibrils of group 1 myocardium tissue illustrates a histogram of random values β, which is somewhat asymmetric with respect to the main extremum ($\beta_0 = 45°$)—Fig. 2.6d.

On the other hand, the ratio between the magnitude of the main extremum and other local extreme values is 100 to 1. Hence there is a high value of the kurtosis of the distribution of the ellipticity of polarisation of the microscopic image of the myocardium sample with ACI. This feature of the two-dimensional distribution $\beta(p \times k)$ can be associated with a certain overwhelming spatial orientation of protein fibrils, which leads to an increase in the probability of the corresponding values of the ellipticity of the polarisation of the laser image. Quantitatively, the features of the morphological construction of the myocardium fibrillar

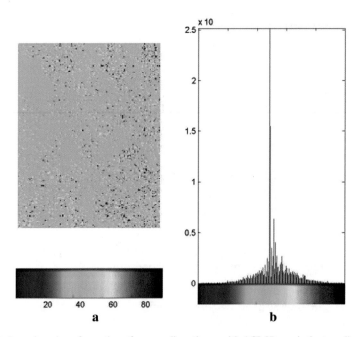

Fig. 2.6 Laser imaging of a section of myocardium tissue with ACI. Here **a** is the two-dimensional ($p \times k$) distribution of polarisation azimuths β; **b** histograms of random polarisation azimuths

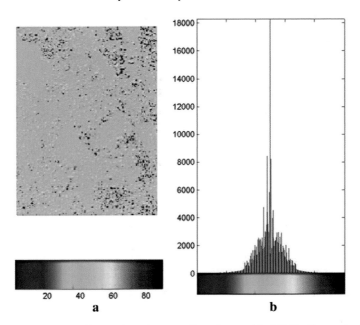

Fig. 2.7 Laser imaging of a section of myocardium tissue with CCHD. Here **a** is the two-dimensional ($p \times k$) distribution of polarisation azimuths β; **b** histograms of random polarisation azimuths

network are revealed in the difference from zero of all the statistical moments: $M_1^\beta = 0.72$; $M_2^\beta = 0.08$; $M_3^\beta = 0.14$; $M_4^\beta = 2.21$.

A polarisation map and a histogram of polarisation ellipticities of a microscopic image of a histological section of myocardium tissue with CCHD are shown in Fig. 2.7.

An analysis of the coordinate structure of the polarisation map (Fig. 2.7a) indicates a slight increase in the birefringence (areas with a difference from the 45° values of β) of the fibrillar network of a group 3 myocardium tissue sample. Quantitatively, such changes are illustrated by the skewness of the histogram of random values β and an increase in its full width-half maximum (Fig. 2.7b). The most distinct statistical changes in the distribution $\beta(p \times k)$ are characterised by an almost 2-fold increase in the value of the statistical moment of the third order and a decrease of 40% in the statistical moment of the fourth order: $M_1^\beta = 0.69$; $M_2^\beta = 0.11$; $M_3^\beta = 0.29$; $M_4^\beta = 1.81$.

The correlation structure of the distribution of polarisation maps $\beta(p \times k)$ of microscopic images of sections of myocardium tissue with ACI and CCHD is illustrated by the autocorrelation functions $K^\beta(\Delta p)$, which are shown in Fig. 2.8.

From the obtained data it can be seen that the coordinate distribution of the polarisation ellipticity values of the microscopic image of the myocardium tissue section with ACI is less uniform than that with CCHD.

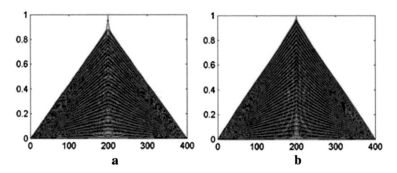

Fig. 2.8 Autocorrelation functions of the distribution of polarisation azimuths β in images of sections of myocardium tissue: **a** ACI; **b** CCHD

This fact is indicated by a slower decrease in the relative values of the autocorrelation function $K^{\beta}(\Delta p)$ calculated for the polarisation map $\beta(p \times k)$ of the microscopic image of a section of myocardium tissue with ACI (Fig. 2.8a) in comparison with the similar correlation dependence that was obtained for myocardium tissue with CCHD (Fig. 2.8b).

Figure 2.9 shows the logarithmic dependences of the power spectra of the distribution of polarisation ellipticity of microscopic images of sections of myocardium tissue at ACI and CCHD.

From the obtained data it can be seen that the distribution of polarisation ellipticity of microscopic images of sections of myocardium tissue with ACI and CCHD is different. For the tissue with ACI, the distribution $\beta(p \times k)$ is fractal, which is confirmed by the constant slope $u = const$ of the approximating curve $A(u)$ of the dependence $\log L(\beta) - \log 1/d$. In the case of CCHD, the distribution $\beta(p \times k)$ of the polarimetric image of the myocardium in the range of small geometrical sizes of

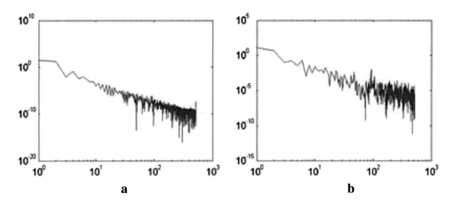

Fig. 2.9 Dependences $\log L(\beta) - \log 1/d$ of microscopic images of myocardium sections: **a** ACI; **b** CCHD

fibrils is statistical, because for approximating curves $A(u)$ there is no stable value
of the angle of inclination $u \neq const$.

2.3   Differential Diagnosis of Myocardium Sections by Azimuthally Invariant Polarisation Mapping of Microscopic Images

This part of the work presents the results of azimuthally stable polarisation mapping
of microscopic images of sections of myocardium of various groups:

- The control group is the intact myocardium (polarisation parameters of samples
  of this type are given in Sect. 2);
- Study group—ACI;
- Comparative group—CCHD.

Figures 2.10 and 2.11 show: polarisation maps of azimuth $\alpha(p \times k)$ and ellipticity
$\beta(p \times k)$; histograms $N(\alpha)$, $N(\beta)$ of the distributions of polarisation parameters of
microscopic images of myocardium sections of ACI (Fig. 2.10) and CCHD (Fig. 2.11)
groups; as well as the corresponding autocorrelation $K(\alpha)$, $K(\beta)$ and $\log L(\alpha) -$
$\log(1/d)$, $\log L(\beta) - \log (1/d)$ dependencies.

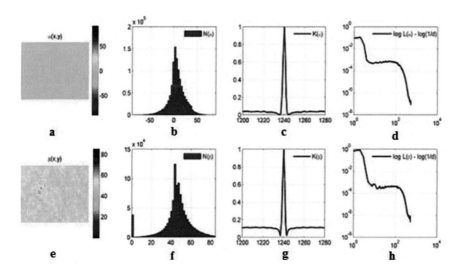

Fig. 2.10 Power spectra of the distribution of polarisation states of microscopic images of sections
of the myocardium that died from ACI. Here **a** polarisation map $\alpha(p \times k)$; **b** distribution histogram
$N(\alpha)$, **c** autocorrelation function $K(\alpha)$; **d** logarithmic dependence of power spectra $\log L(\alpha) -$
$\log (1/d)$; **e** polarisation map $\beta(p \times k)$; **f** distribution histogram $N(\beta)$; **g** autocorrelation function
$K(\beta)$; **h** logarithmic dependence of power spectra $\log L(\beta) - \log (1/d)$

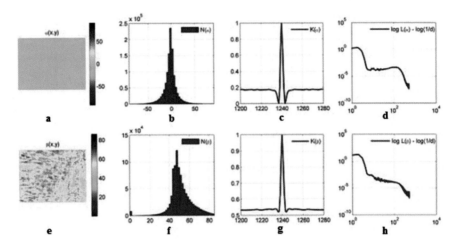

Fig. 2.11 Power spectra of the distribution of polarisation states of microscopic images of sections of the myocardium that died from CCHD. Here **a** polarisation map $\alpha(p \times k)$; **b** distribution histogram $N(\alpha)$, **c** autocorrelation function $K(\alpha)$; **d** logarithmic dependence of power spectra $\log L(\alpha) - \log (1/d)$; **e** polarisation map $\beta(p \times k)$; **f** distribution histogram $N(\beta)$; **g** autocorrelation function $K(\beta)$; **h** logarithmic dependence of power spectra $\log L(\beta) - \log (1/d)$

To determine the statistical reliability of these representative samples of sections, the average deviation σ^2 of the measured values of the statistical and correlation moments was determined by cross-validation.

Statistical processing of the results established a level that provides statistical reliability of the method of azimuthally independent mapping within a representative sample. The resulting deviation corresponds to the value of the confidence interval $p < 0.05$.

Diagnosis of ACI and differentiation with CCHD of myocardium samples for each polarisation map was carried out by cross-comparison of the histograms $N(r)$.

The average $(\bar{r})$ and standard deviation $(\pm\sigma)$ of the parameters r of the polarisation maps $\alpha(p \times k)$ of microscopic images of myocardium samples with ACI and CCHD and the control group are shown in Table 2.1.

Calculations show that the most informative for the diagnosis of ACI are such statistical moments of polarisation maps $\alpha(p \times k)$ as skewness M_3, kurtosis M_4 and kurtosis, M_4^K.

The average $(\bar{r})$ and standard deviation $(\pm\sigma)$ of the parameters r of the polarisation maps $\beta(p \times k)$ of microscopic images of myocardium samples with ACI, CCHD and normal are shown in Table 2.2.

Calculations show that the most informative statistical moments of the parameters r of polarisation maps $\beta(p \times k)$ for the diagnosis of ACI are the same as for polarisation maps $\alpha(p \times k)$, namely skewness M_3, kurtosis M_4 and kurtosis M_4^K.

The following differences between the most sensitive parameters were established for polarisation maps of microscopic images of myocardium sections with ACI and CCHD:

Table 2.1 The average ($\bar{r}$) and standard deviation ($\pm\sigma$) of the parameters r of the polarisation maps $\alpha(p \times k)$ of microscopic images of myocardium samples

Statistical moments	Cause of death		
	Control (n = 20)	CCHD (n = 69)	ACI (n = 69)
Average, M_1	0.16 ± 0.012	0.09 ± 0.008	0.12 ± 0.011
P_1		<0.001	<0.001
P_2		>0.05	
Dispersion, M_2	0.34 ± 0.021	0.21 ± 0.016	0.27 ± 0.017
P_1		<0.001	<0.001
P_2		>0.05	
Skewness, M_3	0.61 ± 0.052	0.39 ± 0.028	0.48 ± 0.045
P_1		<0.001	<0.001
P_2		<0.001	
Kurtosis, M_4	0.13 ± 0.011	0.27 ± 0.023	0.19 ± 0.014
P_1		<0.001	<0.001
P_2		<0.001	
Dispersion, M_2^K	0.07 ± 0.004	0.13 ± 0.011	0.11 ± 0.01
P_1		<0.001	<0.001
P_2		>0.05	
Kurtosis, M_4^K	2.02 ± 0.15	1.23 ± 0.12	1.68 ± 0.15
P_1		<0.001	<0.001
P_2		<0.001	

- $\alpha(p \times k)\begin{cases} \Delta M_1(\alpha) = \Delta M_3(\alpha) = 1.23;\ \Delta M_4(\alpha) = 1.42; \\ \Delta M_4^K(\alpha) = 1.37. \end{cases}$

- $\beta(p \times k)\begin{cases} \Delta M_3(\beta) = 1.27;\ \Delta M_4(\beta) = 1.34; \\ \Delta M_4^K(\beta) = 1.27. \end{cases}$

The results of an analysis of the operational characteristics that describe the diagnostic power of polarisation microscopy are shown in Tables 2.3 and 2.4.

An analysis of the operational characteristics of the polarisation microscopy method, which determine its diagnostic informativeness, found a satisfactory level of such objective parameters as statistical moments of the 1st – 4th order, as well as correlation moments of the 2nd and 4th order characterising the distribution of the azimuth of polarisation of microscopic images in the task of post-mortem ACI diagnostic.

An analysis of the operational characteristics of the polarisation microscopy method, which determines its diagnostic information content in the task of post-mortem ACI diagnostic, revealed a good level of objective parameters such as 3rd and

Table 2.2 The average ($\bar{r}$) and standard deviation ($\pm \sigma$) of the parameters r of the polarisation maps $\beta(p \times k)$ of microscopic images of myocardium samples

Statistical Moments	Cause of death		
	Control (n = 20)	CCHD (n = 69)	ACI (n = 69)
Average, M_1	0.53 ± 0.042	0.61 ± 0.054	0.63 ± 0.059
P_1		<0.001	<0.001
P_2		>0.05	
Dispersion, M_2	0.24 ± 0.017	0.16 ± 0.013	0.19 ± 0.016
P_1		<0.001	<0.001
P_2		>0.05	
Skewness, M_3	0.59 ± 0.044	0.79 ± 0.063	0.62 ± 0.059
P_1		<0.001	<0.001
P_2		<0.001	
Kurtosis, M_4	0.27 ± 0.017	0.47 ± 0.041	0.35 ± 0.029
P_1		<0.001	<0.001
P_2		<0.001	
Dispersion, M_2^K	0.17 ± 0.014	0.11 ± 0.009	0.09 ± 0.008
P_1		<0.001	<0.001
P_2		>0.05	
Kurtosis, M_4^K	1.02 ± 0.094	1.61 ± 0.13	1.27 ± 0.11
P_1		<0.001	<0.001
P_2		<0.001	

Table 2.3 Operational characteristics of multidimensional polarisation microscopy azimuth distributions of images of myocardium sections

R	$Se(\alpha)$, % "1 − 2"	$Sp(\alpha)$, % "1 − 3"	$Se(\alpha)$, % "2 − 3"	$Sp(\alpha)$, % "3 − 2"
Average M_1	67 $a = 46, b = 23$	64 $c = 44, d = 25$	54 $a = 37, b = 32$	52 $c = 36, d = 33$
Dispersion M_2	70 $a = 48, b = 21$	67 $c = 46, d = 23$	57 $a = 39, b = 30$	54 $c = 37, d = 32$
Skewness M_3	78 $a = 54, b = 15$	75 $c = 52, d = 17$	62 $a = 43, b = 26$	59 $c = 41, d = 28$
Kurtosis M_4	79 $a = 56, b = 13$	79 $c = 55, d = 14$	71 $a = 49, b = 20$	64 $c = 44, d = 25$
Average K_2	66 $a = 45, b = 24$	67 $c = 46, d = 23$	58 $a = 40, b = 29$	54 $c = 37, d = 32$
Kurtosis K_4	80 $a = 57, b = 12$	79 $c = 56, d = 13$	75 $a = 52, b = 17$	62 $c = 43, d = 26$

Table 2.4 Operational characteristics of multidimensional polarisation microscopy of the distribution of ellipticity of images of myocardium sections

R	$Se(\alpha), \%$ "1 − 2"	$Sp(\alpha), \%$ "1 − 3"	$Se(\alpha), \%$ "2 − 3"	$Sp(\alpha), \%$ "3 − 2"
Average M_1	70 $a = 48, b = 21$	66 $c = 45, d = 24$	52 $a = 36, b = 33$	51 $c = 35, d = 34$
Dispersion M_2	72 $a = 50, b = 19$	70 $c = 48, d = 21$	53 $a = 37, b = 32$	51 $c = 35, d = 34$
Skewness M_3	83 $a = 57, b = 12$	80 $c = 54, d = 15$	58 $a = 40, b = 29$	55 $c = 38, d = 31$
Kurtosis M_4	86 $a = 59, b = 10$	82 $c = 56, d = 13$	71 $a = 49, b = 23$	62 $c = 43, d = 26$
Average K_2	66 $a = 45, b = 24$	67 $c = 46, d = 23$	58 $a = 40, b = 29$	53 $c = 37, d = 32$
Kurtosis K_4	86 $a = 59, b = 10$	82 $c = 56, d = 13$	72 $a = 50, b = 19$	59 $c = 41, d = 28$

4th order statistical moments, as well as 4th order correlation moments characterising the distribution of polarisation ellipticity of microscopic images.

The results obtained (Tables 3.9 and 3.10) suggest a fairly high level of balanced accuracy, which from the standpoint of evidence-based medicine corresponds to a satisfactory quality of the diagnostic test—$Ac(\beta) = 66.5\%$ and $Ac(\alpha) = 68.5\%$.

The most informative in the differentiation of ACI and CCHD were statistical moments of the 3rd and 4th order, as well as correlation moments of the 4th order:

$$\alpha \Leftrightarrow \begin{cases} Se(\alpha) = 71\% - 75\% \\ Sp(\alpha) = 62\% - 64\% \\ Ac(\alpha) = 67.5\% - 68.5\% \end{cases} ; \beta \Leftrightarrow \begin{cases} Se(\beta) = 71\% - 72\% \\ Sp(\beta) = 59\% - 62\% \\ Ac(\beta) = 65.5\% - 66.5\% \end{cases} .$$

Thus, the method of post-mortem diagnosis of myocardium based on azimuthally invariant polarisation mapping of microscopic images is presented. It has an adequate analytical justification and good reproducibility.

2.4 Post-mortem Diagnosis of Acute Coronary Myocardium Insufficiency Using Wavelet Analysis of Azimuthally Invariant Polarisation Maps of Microscopic Images

This section presents the results of studies of ACI and CCHD cases by using wavelet analysis of the azimuth coordinate distributions and polarisation ellipticity of microscopic images of myocardium slices in order to improve the balanced accuracy of the diagnostic method.

As already noted, wavelet analysis allows one to separately evaluate the manifestations of the optical properties of small-scale polypeptide chains of optically active myosin molecules (polarisation azimuth maps) and large-scale fibrillar networks (polarisation ellipticity maps) of myocardium slices.

ACI, as a manifestation of acute ischemia, is characterised by massive shifts during biochemical processes in the cell. First of all, there is a transition to the aerobic mechanism of glycolysis with the accumulation of reducing equivalents and simultaneous inhibition of energy-dependent processes—the functioning of the actin-myosin complex and ion pumps in combination with the launch of lipid peroxidation. These changes lead to the formation of actinomyosin bridges that underlie contractual damage to the contractile apparatus.

Based on this, it can be assumed that the greatest differences between the normal myocardium and myocardium in the case of ACI should be sought at small scales of change in the structure of the polypeptide chains of myocardium proteins.

In contrast, the major optical manifestations of CCHD are in the transformation of large-scale fibrillar myosin networks. Therefore, a scale-selective approach to the analysis of polarisation maps of microscopic images of sections of both groups will increase the accuracy of the posthumous diagnosis of ACI.

The results of studies of the wavelet-coefficient distributions characterising the manifestations of the optical properties of anisotropic myocardium networks at different scales of polarisation maps of microscopic images of myocardium sections at ACI and CCHD are shown in the series of Figs. 2.12, 2.13, 2.14 and 2.15.

Figures 2.12, 2.13, 2.14 and 2.15. contain three main fragments:

- polarisation maps $\{\alpha(x, y); \beta(x, y)\}$ of microscopic images of myocardium samples that died due to ACI and CCHD;
- two-dimensional distributions of wavelet-coefficients $W_{a,b}\{\alpha(x, y); \beta(x, y)\}$ characterising the structure of polarisation maps formed by different-scale morphological components of fibrillar networks;
- linear sections $C_{a=15,b}$ and $C_{a=55,b}$ of the wavelet-maps $W_{a,b}$ on the scale of the wavelet-function $a = 15$ and $a = 55$, which provides a separate assessment of changes in optical anisotropy at the level of individual polypeptide chains ($a = 15$) and the levels of their organisation into fibrils ($a = 55$).

A comparative analysis of the results of the study of the polarisation azimuth of the coordinate, multiscale distributions of wavelet-coefficients $W_{a,b}(\alpha)$ and $W_{a,b}(\beta)$, characterising polarisation-inhomogeneous microscopic images of myocardium tissue samples which died due to ACI and CCHD revealed the greatest differences between them at the level of small scale a_{min} distributions of wavelet-coefficients, as indicated by the modulation of "small-scale "dependencies $C_{a=15,b}$ of such wavelet-cards.

The most clearly established changes were manifested for the polarisation map of the azimuth of images of native myocardium sections (Figs. 2.12 and 2.13).

The features established are related to the fact that ischemic changes in polycrystalline myocardium fibrillar networks occur not at the morphological (large-scale), but at the concentration (small-scale) levels of its morphological structure.

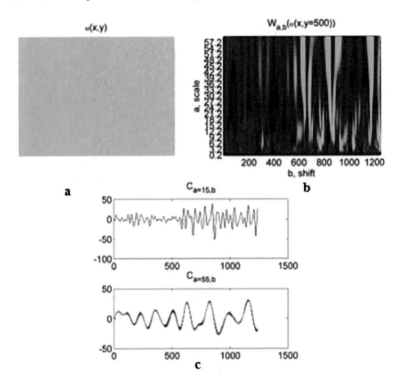

Fig. 2.12 Distribution of the wavelet coefficient of the polarisation map of the azimuth of polarisation of the microscopic image of a myocardium section at ACI. Here **a** is the two-dimensional distribution of the polarisation azimuth; **b** two-dimensional distribution of wavelet-coefficients $W_{a,b}(\alpha)$; **c** sections of wavelet-coefficients on the scales $C_{a=15,b}$ and $C_{a=55,b}$

Due to a change in the concentration of optically active myosin molecules due to the occurrence of ischemic changes due to ACI, the level of optical activity decreases, which manifests itself in the formation of rotations of the plane of polarisation of the points of the corresponding microscopic image of the native section. In accordance with this, the depth of modulation of the amplitudes of wavelet-coefficients at small a_{min} scales of the azimuthally-invariant map of polarisation azimuths $\alpha(x, y)$ also decreases.

A comparative analysis of the results of the study of the polarisation ellipticity of the coordinate, multiscale distributions of wavelet-coefficients $W_{a,b}(\alpha)$ and $W_{a,b}(\beta)$ (Figs. 2.14 and 2.15), which characterise polarisation-inhomogeneous microscopic images of myocardium tissue samples deceased due to ACI and CCHD, revealed the largest differences between them at the level of large scales a_{max} distribution of wavelet-coefficients, as indicated by the modulation of the "large-scale" dependencies $C_{a=55,b}$ of such wavelet cards.

By means of statistical analysis, it was found that a quantitative decrease in the modulation depth of the amplitudes of wavelet-coefficients at small a_{min} scales of

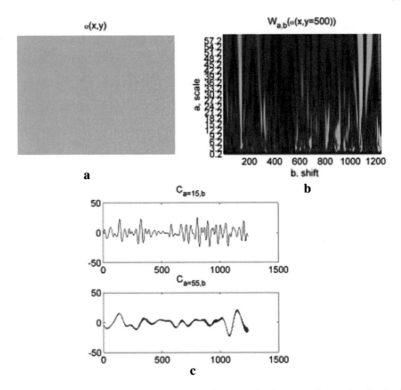

Fig. 2.13 Distribution of the wavelet coefficient of the polarisation map of the azimuth of polarisation of the microscopic image of a myocardium section at CCHD. Here **a** is the two-dimensional distribution of the polarisation azimuth; **b** two-dimensional distribution of wavelet-coefficients $W_{a,b}(\alpha)$; **c** sections of wavelet-coefficients on the scales $C_{a=15,b}$ and $C_{a=55,b}$

the azimuthally-invariant map of polarisation azimuths $\alpha(x, y)$ turns out to give a decrease in dispersion $M_{i=2}\left(C_{a=15,b}(\alpha)\right)\downarrow$ and a corresponding increase in the values of statistical moments of higher orders $M_{i=3;4}\left(C_{a=15,b}(\alpha)\right)\uparrow$ characterising the distribution of wavelet-coefficients (Table 2.5.).

In particular, at the scale $a_{\min}$, statistical moments M_2, M_3, M_4 were the most sensitive to ACI, and at the scale $a_{\max}$, M_3, M_4 were most sensitive to CCHD.

Based on the data obtained, quantitative parameters of the difference between the statistical moments that characterise the distribution of wavelet-coefficients at different scales of the polarisation map of the azimuth of microscopic images for statistically significant differentiation of myocardium samples deceased due to ACI and CCHD are determined:

$$\alpha(p \times k) \Rightarrow \begin{Bmatrix} a_{234_{\min}} \\ a_{34_{\max}} \end{Bmatrix}.$$

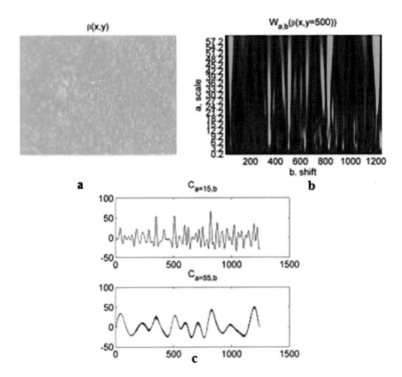

Fig. 2.14 Distribution of the wavelet-coefficients of the polarisation map of the ellipticity of polarisation of the microscopic image of a myocardium section at ACI. Here **a** is the two-dimensional distribution of the polarisation azimuth; **b** two-dimensional distribution of wavelet-coefficients $W_{a,b}(\beta)$; **c** sections of wavelet-coefficients on the scales $C_{a=15,b}$ and $C_{a=55,b}$

On the other hand, the wavelet analysis of polarisation maps of the ellipticity of points of digital microscopic images of myocardium sections both with ACI and with CCHD turned out to be more sensitive to ischemic changes in the myocardium on large scales of the fibrillar network (Table 2.6).

An analysis of the data presented in the table shows that the most sensitive parameters to ischemic changes on a large-scale level of the myocardium structure were higher-order statistical moments, characterising the distribution of wavelet-coefficients on a large scale $a_{max} = 55$ of the polarisation map of ellipticity of points of microscopic images, namely M_3 and M_4.

Quantitative parameters of the difference between the indicated objective parameters were determined for statistically reliable verification of myocardium samples deceased due to ACI:

$$\alpha(p \times k) \Rightarrow \{a_{34_{max}}\}.$$

The results of the analysis of the operational characteristics characterising the diagnostic power of the post-mortem myocardium diagnosis by wavelet analysis of

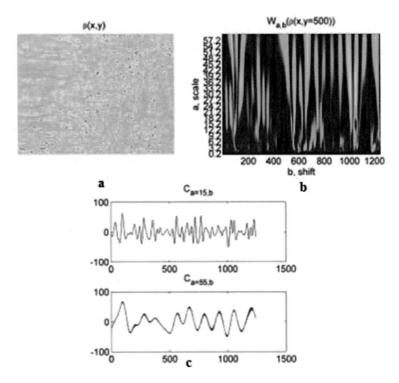

Fig. 2.15 Distribution of the wavelet-coefficients of the polarisation map of the ellipticity of polarisation of the microscopic image of a myocardium section at CCHD. Here **a** is the two-dimensional distribution of the polarisation azimuth; **b** two-dimensional distribution of wavelet-coefficients $W_{a,b}(\alpha)$; **c** sections of wavelet-coefficients on the scales $C_{a=15,b}$ and $C_{a=55,b}$

polarisation maps of azimuth and ellipticity of microscopic images of native sections of the myocardium deceased due to ACI and CCHD at small (a_{min})—Tables 2.7 and 2.8—and large ($a_{max} = 55$)—Tables 2.9 and 2.10—scales are given.

An analysis of the operational characteristics of the scale-selective method of wavelet-analysis of polarisation microscopy data, which determine its diagnostic information content, revealed an increase (in comparison with direct analysis of polarisation maps) of the balanced accuracy of the post-mortem myocardium diagnosis.

On large-scale (a_{max}) assessment of manifestations of optical anisotropy of fibrillar networks, the third-order statistical moment was the most informative, characterising the skewness of the distribution of the wavelet-coefficients of the polarisation ellipticity map of the microscopic images of myocardium sections at ACI and CCHD

Table 2.5 Statistical moments of the first to fourth orders of magnitude characterising the distribution $C_{a=15,b}$ and $C_{a-55,b}$ wavelet-coefficients of the map of the azimuths of polarisation of microscopic images of myocardium sections

Statistical moments	Cause of death		
	Control (n = 20)	CCHD (n = 69)	ACI (n = 69)
a_{min}			
Average, M_1	0.009 ± 0.001	0.013 ± 0.0011	0.015 ± 0.0012
P_1		<0.001	<0.001
P_2		>0.05	
Dispersion, M_2	0.094 ± 0.01	0.18 ± 0.014	0.13 ± 0.011
P_1		<0.001	<0.001
P_2		<0.001	
Skewness, M_3	0.16 ± 0.012	0.38 ± 0.026	0.25 ± 0.022
P_1		<0.001	<0.001
P_2		<0.001	
Kurtosis, M_4	1.13 ± 0.11	0.67 ± 0.049	0.89 ± 0.091
P_1		<0.001	<0.001
P_2		<0.001	
a_{max}			
Average, M_1	0.05 ± 0.002	0.06 ± 0.004	0.07 ± 0.006
P_1		>0.05	>0.05
P_2		>0.05	
Dispersion, M_2	0.28 ± 0.023	0.23 ± 0.021	0.25 ± 0.022
P_1		>0.05	>0.05
P_2		>0.05	
Skewness, M_3	0.31 ± 0.025	0.19 ± 0.018	0.24 ± 0.018
P_1		<0.001	<0.001
P_2		<0.001	
Kurtosis, M_4	1.08 ± 0.097	0.72 ± 0.068	0.86 ± 0.071
P_1		<0.001	<0.001
P_2		<0.001	

$$\beta \Leftrightarrow \begin{cases} Se(\beta) = 62\% - 78\% \\ Sp(\beta) = 57\% - 67\% \\ Ac(\beta) = 59.5\% - 72.5\% \end{cases}.$$

On a small scale (a_{min}) assessment of manifestations of the optical activity of the primary structure of myosin proteins, the third and fourth order statistical moments were the most informative, characterising the distribution of the wavelet-coefficients of the polarisation azimuth maps of microscopic images of native myocardium sections deceased due to ACI and CCHD

Table 2.6 5 Statistical moments of the first to fourth orders of magnitude characterising the distribution $C_{a=15,b}$ and $C_{a=55,b}$ wavelet-coefficients of the map of the ellipticity of polarisation of microscopic images of myocardium sections

Statistical moments	Cause of death		
	Control (n = 20)	CCHD (n = 69)	ACI (n = 69)
a_{min}			
Average, M_1	0.027 ± 0.0021	0.031 ± 0.024	0.029 ± 0.0022
P_1		>0.05	>0.05
P_2		>0.05	
Dispersion, M_2	0.1 ± 0.009	0.13 ± 0.011	0.11 ± 0.008
P_1		>0.05	>0.05
P_2		>0.05	
Skewness, M_3	0.44 ± 0.039	0.53 ± 0.042	0.45 ± 0.041
P_1		>0.05	>0.05
P_2		>0.05	
Kurtosis, M_4	0.83 ± 0.071	0.75 ± 0.068	0.79 ± 0.064
P_1		>0.05	>0.05
P_2		>0.05	
a_{max}			
Average, M_1	0.14 ± 0.009	0.11 ± 0.007	0.12 ± 0.008
P_1		>0.05	>0.05
P_2		>0.05	
Dispersion, M_2	0.29 ± 0.024	0.25 ± 0.021	0.28 ± 0.023
P_1		>0.05	>0.05
P_2		>0.05	
Skewness, M_3	0.81 ± 0.075	0.49 ± 0.038	0.63 ± 0.058
P_1		<0.001	<0.001
P_2		<0.001	
Kurtosis, M_4	1.36 ± 0.11	0.91 ± 0.082	1.18 ± 0.11
P_1		<0.001	<0.001
P_2		<0.001	

$$\alpha \Leftrightarrow \begin{cases} Se(\alpha) = 64\% - 77\% \\ Sp(\alpha) = 59\% - 68\% \\ Ac(\alpha) = 61.5\% - 72.5\% \end{cases}.$$

The obtained results on the operational characteristics of the polarisation microscopy method with scale-selective analysis (Tables 2.9 and 2.10) revealed an increase in balanced accuracy to the level of a good quality diagnostic test—$Ac(\beta) = 72.5\%$ and $Ac(\alpha) = 72.5\%$.

Table 2.7 Operational characteristics of the wavelet-analysis of the azimuth distribution of images of myocardium sections on a scale a_{min}

R	$Se(\alpha), \%$ "1 − 2"	$Sp(\alpha), \%$ "1 − 3"	$Se(\alpha), \%$ "2 − 3"	$Sp(\alpha), \%$ "3 − 2"
Average M_1	70 $a = 48, b = 21$	67 $c = 46, d = 23$	54 $a = 37, b = 32$	52 $c = 36, d = 33$
Dispersion M_2	74 $a = 51, b = 18$	70 $c = 48, d = 21$	59 $a = 41, b = 28$	55 $c = 38, d = 31$
Skewness M_3	78 $a = 54, b = 15$	74 $a = 51, b = 18$	64 $a = 41, b = 28$	59 $c = 38, d = 31$
Kurtosis M_4	88 $a = 61, b = 8$	78 $a = 54, b = 15$	77 $a = 53, b = 16$	68 $c = 47, d = 22$

Table 2.8 Operational characteristics of the wavelet-analysis of the ellipticity distribution of images of myocardium sections on a scale a_{min}

R	$Se(\alpha), \%$ "1 − 2"	$Sp(\alpha), \%$ "1 − 3"	$Se(\alpha), \%$ "2 − 3"	$Sp(\alpha), \%$ "3 − 2"
Average M_1	57 $a = 39, b = 30$	56 $c = 38, d = 31$	54 $a = 37, b = 32$	52 $c = 36, d = 33$
Dispersion M_2	60 $a = 41, b = 28$	51 $c = 40, d = 29$	57 $a = 39, b = 30$	51 $c = 35, d = 34$
Skewness M_3	65 $a = 45, b = 24$	63 $c = 44, d = 25$	62 $a = 43, b = 26$	58 $c = 40, d = 29$
Kurtosis M_4	76 $a = 52, b = 17$	74 $c = 50, d = 19$	71 $a = 49, b = 20$	65 $c = 45, d = 24$

Table 2.9 Operational characteristics of the wavelet-analysis of the azimuth distribution of images of myocardium sections on a scale a_{max}

R	$Se(\alpha), \%$ "1 − 2"	$Sp(\alpha), \%$ "1 − 3"	$Se(\alpha), \%$ "2 − 3"	$Sp(\alpha), \%$ "3 − 2"
Average M_1	62 $a = 43, b = 26$	61 $c = 42, d = 27$	57 $a = 39, b = 30$	57 $c = 39, d = 30$
Dispersion M_2	67 $a = 46, b = 23$	65 $c = 45, d = 24$	62 $a = 43, b = 26$	57 $c = 39, d = 30$
Skewness M_3	77 $a = 54, b = 15$	75 $c = 52, d = 17$	67 $a = 46, b = 23$	65 $c = 45, d = 24$
Kurtosis M_4	83 $a = 57, b = 12$	81 $c = 55, d = 14$	75 $a = 52, b = 17$	71 $c = 49, d = 20$

Thus, the presented method of post-mortem myocardium diagnosis based on azimuthally-invariant polarising mapping of microscopic images has an adequate analytical justification and corresponding reproducibility.

In parallel with this, the method of wavelet-analysis of the distribution of polarisation states of microscopic images of myocardium sections expands the functionality of azimuthally-invariant polarisation mapping.

Table 2.10 Operational characteristics of the wavelet-analysis of the ellipticity distribution of images of myocardium sections on a scale a_{max}

R	$Se(\alpha), \% \text{ "1}-2\text{"}$	$Sp(\alpha), \% \text{ "1}-3\text{"}$	$Se(\alpha), \% \text{ "2}-3\text{"}$	$Sp(\alpha), \% \text{ "3}-2\text{"}$
Average M_1	57 $a = 39, b = 30$	56 $c = 38, d = 31$	54 $a = 37, b = 32$	52 $c = 36, d = 33$
Dispersion M_2	60 $a = 41, b = 28$	51 $c = 40, d = 29$	57 $a = 39, b = 30$	51 $c = 35, d = 34$
Skewness M_3	77 $a = 53, b = 16$	72 $c = 50, d = 21$	62 $a = 43, b = 26$	58 $c = 40, d = 29$
Kurtosis M_4	83 $a = 57, b = 12$	78 $c = 54, d = 15$	71 $a = 49, b = 20$	65 $c = 45, d = 24$

2.5 Conclusions

Testing the method of azimuthally-invariant polarisation mapping of post-mortem changes in the optical anisotropy of the sections of the myocardium of both groups provided a satisfactory level of balanced accuracy—$Ac \sim 65\% - 68\%$ for the detection of ACI and differentiation from CCHD.

1. The possibilities of using wavelet analysis of azimuthally-independent polarisation maps $r = \{\alpha(x, y); \beta(x, y)\}$ of microscopic images of sections in the post-mortem diagnosis of myocardium ACI are considered.

2. The wavelet-analysis of polarisation maps of microscopic images of the optical manifestations of the primary and secondary structures of the myocardium fibrillar network revealed the sensitivity of a set of statistical moments of the first and fourth orders characterising the distribution of the amplitudes of the wavelet-coefficients at different scales up to post-mortem changes in the myocardium. On this basis, the post-mortem diagnosis of myocardium ACI with good balanced accuracy was first implemented. $Ac = 72.5\%–73\%$.

References

1. A. Ushenko, V. Pishak, Laser polarimetry of biological tissue: principles and applications, in *Handbook of Coherent-Domain Optical Methods: Biomedical Diagnostics*, ed. by V. Tuchin (Environmental and Material Science, 2004), pp. 93–138
2. O. Angelsky, A. Ushenko, Y. Ushenko, V. Pishak, A. Peresunko, Statistical, correlation and topological approaches in diagnostics of the structure and physiological state of birefringent biological tissues. Handbook Photonics Biomed. Sci. (2010), pp. 283–322
3. Y. Ushenko, T. Boychuk, V. Bachynsky, O. Mincer, Diagnostics of structure and physiological state of birefringent biological tissues: statistical, correlation and topological approaches, in *Handbook of Coherent-Domain Optical Methods,* ed. by V. Tuchin (Springer Science+Business Media; 2013)

4. A.G. Ushenko, A.V. Dubolazov, V.A. Ushenko, Yu.A. Ushenko, M. Yu. Sakhnovskiy, O.I. Olar, Methods and means of laser polarimetry microscopy of optically anisotropic biological layers. Proc. SPIE Int. Soc. Opt. Eng. **9971**(99712B) (2016)
5. M.Y. Sakhnovskiy, A.-V. Syvokorovskaya, V. Martseniak, B.M. Bodnar, O. Tsyhykalo, A.V. Dubolazov, O.I. Olar, V.A. Ushenko, P.M. Grygoryshyn, System of biological crystals fibrillar networks polarisation-correlation mapping. Proc. SPIE Int. Soc. Opt. Eng. **10752**, 107522G (2018)
6. M.Y. Sakhnovskiy, O.Y. Wanchuliak, B. Bodnar, I.V. Martseniak, O. Tsyhykalo, A.V. Dubolazov, V.A. Ushenko, O.I. Olar, P.M. Grygoryshyn, Polarisation-interference images of optically anisotropic biological layers. Proc. SPIE Int. Soc. Opt. Eng. **10752**, 107522F (2018)
7. Y.A. Ushenko, O.V. Olar, A.V. Dubolazov, O.B. Bodnar, B.M. Bodnar, L. Pidkamin, O. Prydiy, M.I. Sidor, D. Kvasnyuk, O. Tsyhykalo, System of differential mueller matrix mapping of phase and amplitude anisotropy of depolarising biological tissues. Proc. SPIE Int Soc Opt Eng **10752**, 107522H (2018)
8. N. Sivokorovskaya, V.T. Bachinskyi, O.Y. Vanchulyak, O.G. Ushenko, A.V. Dubolazov, Y.O. Ushenko, Y.Y. Tomka, L.Y. Kushnerik, Statistical analysis of polarisation images of histological cuts of parenchymatic tissues in diagnostics of volume of blood loss. IFMBE Proc. **77**, 513–517 (2020)
9. O.G. Ushenko, A.-V. Syvokorovskaya, V.T. Bachinsky, O.Y. Vanchuliak, A.V. Dubolazov, Y.O. Ushenko, Y.Y. Tomka, M.L. Kovalchuk, Laser autofluorescent microscopy of histological sections of parenchymatous biological tissues of the dead. IFMBE Proc. **77**, 507–511
10. Y. Sarkisova, V.T. Bachinskyi, M. Garazdyuk, O.Y. Vanchulyak, O.Y. Litvinenko, O.G. Ushenko, B.G. Bodnar, A.V. Dubolazov, Y.O. Ushenko, Y.Y. Tomka, I.V. Soltys, S. Foglinskiy, Differential muller-matrix microscopy of protein fractions of vitreous preparations in diagnostics of the pressure of death. IFMBE Proc. **77**, 503–506 (2020)
11. D.P. Zipes, A.J. Camm, M. Borggrefe, et al., ACC/AHA/ESC 2006 guidelines for management of patients with ventricular arrhythmias and the prevention of sudden cardiac death: a report of the American College of Cardiology/American Heart association task force and the european society of cardiology committee for practice guidelines (writing committee to develop guidelines for management of patients with ventricular arrhythmias and the prevention of sudden cardiac death). J. Am. Coll. Cardiol. **48**(5), e247-e346 (2006)
12. S.V. De Noronha, S. Sharma, M. Papadakis et al., Aetiology of sudden cardiac death in athletes in the United Kingdom: a pathological study. Heart **95**(17), 1409–1414 (2009)
13. Agrafioti, F., Gao, J., Hatzinakos, D.: Heart biometrics: theory, methods and applications. Biometrics: Book, **3**, 199–216 (2011)
14. American Heart Association, *Older Americans and cardiovascular diseases—statistics dallas* (American Heart Association, Texas, 2005)
15. Hind Al-Khayat, A., Robert Kensler, W., S. John, et al., Anatomic model of the human cardiac muscle myosin filament. Proct. Nat. Acad Sci USA, **110**(1), 318–323 (2013)
16. S. Muller-Nordhorn, S. Binting, S. Roll, S.N. Willich, An update on regional variation in cardiovascular mortality within Europe. Eur. Heart J. **29**, 1316–1326 (2008)
17. S. Nattel, A. Maguy, S. Bouter, Yeh Y.H. Le, Arrhythmogenic ion-channel remodeling in the heart: heart failure, myocardium infarction, and atrial fibrillation. Physiol. Rev. **87**(2), 425–456 (2007)
18. J.T. Tikkanen, V. Wichmann, M.J. Junttila, et al., Association of early repolarisation and sudden cardiac death during an acute coronary event. Circ. Arrhythmia Electrophysiol. **5**(4), 714–718 (2012)
19. N. Franceschini, C. Carty, P. Buzková, Reiner association of genetic variants and incident coronary heart disease in multiethnic cohorts: the PAGE study. Circ. Cardiovasc. Genet **4**(6), 661–672 (2011)
20. V. Tuchin, L. Wang, D. Zimnjakov, *Optical Polarisation in Biomedical Applications* (Springer, New York, USA, 2006)
21. Polarimetry, C.R., Handbook of optic: Vol I— geometrical and physical optics, polarised light, components and instruments, in ed. by M. Bass (McGraw-Hill Professional, New York 2010), pp. 22.1–22.37

22. N. Ghosh, M. Wood, A. Vitkin, Polarised light assessment of complex turbid media such as biological tissues via Mueller matrix decomposition, in *Handbook of Photonics for Biomedical Science,* ed. by V. Tuchin (CRC Press, Taylor & Francis Group; London, 2010). pp. 253–282

23. S. Jacques, Polarised light imaging of biological tissues, in *Handbook of Biomedical Optics,* ed. by D. Boas, C. Pitris, N. Ramanujam (CRC Press, Boca Raton, London, New York, 2011), pp. 649–669

24. N. Ghosh, Tissue polarimetry: concepts, challenges, applications, and outlook. J. Biomed. Opt. **16**(11), (2011)

25. M. Swami, H. Patel, P. Gupta, Conversion of 3×3 Mueller matrix to 4×4 Mueller matrix for non-depolarising samples. Opt. Commun. **286**, 18–22 (2013)

26. Y. Ushenko, V. Ushenko, A. Dubolazov, V. Balanetskaya, N. Zabolotna, Mueller matrix diagnostics of optical properties of polycrystalline networks of human blood plasma. Opt. Spectrosc. **112**(6), 884–892 (2012)

27. V. Ushenko, O. Dubolazov, A. Karachevtsev, Two wavelength Mueller matrix reconstruction of blood plasma films polycrystalline structure in diagnostics of breast cancer. Appl. Opt. **53**(10), B128 (2016)

28. Y. Ushenko, G. Koval, A. Ushenko, O. Dubolazov, V. Ushenko, O. Novakovskaia, Mueller matrix of laser-induced autofluorescence of polycrystalline films of dried peritoneal fluid in diagnostics of endometriosis. J. Biomed. Opt. **21**(7), (2016)

29. A.G. Ushenko, A.V. Dubolazov, V.A. Ushenko, Y.A. Ushenko, L.Ya. Kushnerick, O.V. Olar, N.V. Pashkovskaya, Y.F. Marchuk, Mueller matrix differentiation of fibrillar networks of biological tissues with different phase and amplitude anisotropy. Proc. SPIE Int. Soc. Opt. Eng. **9971**, 99712 K (2016)

30. O.V. Dubolazov, A.G. Ushenko, Y.A. Ushenko, M.Y. Sakhnovskiy, P.M. Grygoryshyn, N. Pavlyukovich, O.V. Pavlyukovich, V.T. Bachynskiy, S.V. Pavlov, V.D. Mishalov, Z. Omiotek, O. Mamyrbaev, Laser müller matrix diagnostics of changes in the optical anisotropy of biological tissues, in *Information Technology in Medical Diagnostics II—Proceedings of the International Scientific Internet Conference on Computer Graphics and Image Processing and 48th International Scientific and Practical Conference on Application of Lasers in Medicine and Biology, 2018* (2019), pp. 195–203

31. V.A. Ushenko, A.Y. Sdobnov, W.D. Mishalov, A.V. Dubolazov, O.V. Olar, V.T. Bachinskyi, A.G. Ushenko, Y.A. Ushenko, O.Y. Wanchuliak, I. Meglinski, Biomedical applications of jones-matrix tomography to polycrystalline films of biological fluids. J. Innovat. Opt. Health Sci. **12** (6), 1950017 (2019)

32. M. Borovkova, L. Trifonyuk, V. Ushenko, O. Dubolazov, O. Vanchulyak, G. Bodnar, Y. Ushenko, O. Olar, O. Ushenko, M. Sakhnovskiy, A. Bykov, I. Meglinski, Mueller matrix-based polarisation imaging and quantitative assessment of optically anisotropic polycrystalline networks. PLoS ONE, **14**(5), e0214494 (2019)

33. D. Layden, N. Ghosh, A. Vitkin, Quantitative polarimetry for tissue characterisation and diagnosis, in *Advanced Biophotonics: Tissue Optical Sectioning,* ed. by R. Wang, V. Tuchin (CRC Press, Taylor & Francis Group, Boca Raton, London, New York, 2013), pp. 73–108

34. T. Vo-Dinh, *Biomedical Photonics Handbook: 3 volume Set.* 2nd ed. (CRC Press, Boca Raton, 2014)

35. A. Vitkin, N. Ghosh, A. Martino, Tissue polarimetry, in *Photonics: scientific foundations, technology and applications,* 4th edn., ed. by D. Andrews (Wiley, Hoboken, New Jersey, 2015), pp. 239–321

36. V. Tuchin, *Tissue Optics: Light Scattering Methods and Instruments for Medical Diagnosis,* 2nd edn. (SPIE Press, Bellingham, Washington, USA, 2007)

37. W. Bickel, W. Bailey, Stokes vectors, Mueller matrices, and polarised scattered light. Am. J. Phys. **53**(5), 468–478 (1985)

38. A. Doronin, C. Macdonald, I. Meglinski, Propagation of coherent polarised light in turbid highly scattering medium. J. Biomed. Opt. **19**(2), (2014)

39. A. Doronin, A. Radosevich, V. Backman, I. Meglinski, Two electric field Monte Carlo models of coherent backscattering of polarised light. J. Opt. Soc. Am. A. **31**(11), 2394 (2014)

40. P.A. Gopinathan, G. Kokila, M. Jyothi, C. Ananjan, L. Pradeep, S.H. Nazir, Study of collagen birefringence in different grades of oral squamous cell carcinoma using picrosirius red and polarised light microscopy. Scientifica **2015**, 802980 (2015)
41. L. Rich, P. Whittaker, Collagen and picrosirius red staining: a polarised light assessment of fibrillar hue and spatial distriburuon. Braz. J. Morphol. Sci. (2005)
42. S. Bancelin, A. Nazac, B.H. Ibrahim, et al., Determination of collagen fiber orientation in histological slides using Mueller microscopy and validation by second harmonic generation imaging. Opt. Express, **22**(19), 22561–22574 (2014)
43. O. Angelsky, Y. Tomka, A. Ushenko, Y. Ushenko, S. Yermolenko, 2-D tomography of biotissue images in pre-clinic diagnostics of their pre-cancer states. Proc. SPIE **5972**, 158–162 (2005)

Chapter 3
Diagnosis of Acute Coronary Insufficiency by the Method of Mueller Matrix Analysis of Myosin Myocardium Networks

The addition and development of laser polarimetry of azimuth and ellipticity distribution of microscopic images of biological layers (BL) is a combination of methods from polarisation nephelometry [1]. These methods are based on determining parameters of microscopic images that are statistically averaged over the entire set of optical inhomogeneities. The achievement of statistical averaging is based on the determination of the elements of the light scattering matrix (Mueller matrix) [2–4], which carries the most complete information on the polarisation properties of biological objects.

Thus, measurements of quantities that describe the polarisation vector (Stokes vector) of the radiation converted by BL make it possible to obtain the most complete (statistically averaged for all inhomogeneities of a biological object) information on the polarisation properties of the latter [1, 5–22].

Three important groups can be distinguished within the elements of the Mueller matrix $m_{ik}(p \times k)$ of myocardium tissue:

- coordinate distribution of the values of the elements of the Mueller matrix $m_{22;33}(p \times k)$. Such elements characterise the degree of conversion of the azimuth of the polarisation of the laser wave by myosin fibrils, the optical axes of which are oriented in two mutually perpendicular directions $\gamma = 0° \leftrightarrow 90°$ ($m_{22}(p \times k)$) and $\gamma = 45° \leftrightarrow 135°$ ($m_{33}(p \times k)$), respectively;
- coordinate distributions of the matrix element $m_{44}(p \times k)$. The value of this element is determined by the phase shifts between the orthogonal components of the amplitude of the laser wave arising due to the birefringence of the substance of myosin fibrils;
- coordinate distributions of the off-diagonal elements of the Mueller matrix $m_{23;24;34}(p \times k)$, characterising the mechanisms of mutual transformations of a linear polarisation, via the myocardium fibrillar network, into elliptical and vice versa.

V. Bachinsky et al., *Multi-parameter Mueller Matrix Microscopy for the Expert Assessment of Acute Myocardium Ischemia*, SpringerBriefs in Applied Sciences and Technology, https://doi.org/10.1007/978-981-16-1450-7_3

This section explores the possibilities of diagnosing the onset of death as a result of ACI by determining the statistical moments of the first to fourth orders of magnitude that characterise the distribution of the elements of the Mueller matrix of myocardium tissue sections.

3.1 Diagnostic Capabilities of Multi-Parameter Analysis of the Distributions of the "Orientation" Element of the Myocardium Tissue of the Mueller Matrix

Statistical Analysis

The results of experimental studies of the coordinate $m_{22}(p \times k)$ and statistical $N(m_{22})$ structure of the "orientational" elements m_{22} of the Mueller matrix of myocardium sections with ACI and CCHD are illustrated in Fig. 3.1.

It can be seen from the obtained data that the use of the coordinate structure of the "orientation" element m_{22} of the myocardium Mueller matrix for visual establishment of ACI or for differentiation with other states (CCHD) is difficult.

The distributions $m_{22}(p \times k)$ characterising the distribution of the directions of the optical axes of myosin myocardium fibrils are coordinate-heterogeneous and have the same range of eigenvalues $-1 \leq m_{22} \leq 1$.

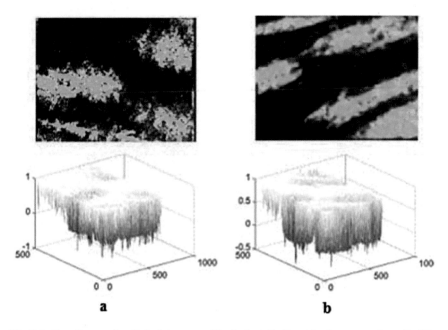

Fig. 3.1 Coordinate and statistical structure of the "orientation" matrix element $m_{22}(p \times k)$ of the myocardium of both groups: **a** ACI group; **b** CCHD group

Table 3.1 Statistical moments $M_j(m_{22})$ that characterise the coordinate distribution $m_{22}(p \times k)$ of myocardium tissue

Statistical moments	Cause of death		
	Control (n = 20)	CCHD (n = 69)	ACI (n = 69)
Average, M_1	0.89 ± 0.072	0.72 ± 0.058	0.86 ± 0.067
P_1		>0.05	>0.05
P_2		>0.05	
Dispersion, M_2	0.12 ± 0.01	0.13 ± 0.011	0.11 ± 0.012
P_1		>0.05	>0.05
P_2		>0.05	
Skewness, M_3	0.29 ± 0.023	0.24 ± 0.019	0.28 ± 0.026
P_1		>0.05	>0.05
P_2		>0.05	
Kurtosis, M_4	0.58 ± 0.047	0.51 ± 0.046	0.67 ± 0.057
P_1		>0.05	>0.05
P_2		>0.05	

Quantitatively, the structure of the distributions of phase elements Z_{22} defined for two groups of samples (CCHD—$n = 69$ and ACI—$n = 69$) of the myocardium is illustrated by the magnitudes and ranges of values (Table 3.1).

A comparative analysis of the values and ranges of variation of the values of statistical moments of the 1st and 4th orders $M_{j=1;2;3;4}(m_{22})$, which characterise the coordinate distributions of the matrix elements describing the orientational structure of myocardium fibrillar networks within the studied groups, did not reveal an objective (statistically significant) possibility of ACI verification, since the ranges of variation of the average $M_1(m_{22})$, dispersion $M_2(m_{22})$, skewness $M_3(m_{22})$ and kurtosis $M_4(m_{22})$ values overlap for different groups of myocardium sections.

In order to search for more sensitive diagnostic criteria, a parametric analysis of Mueller matrix images was applied—the statistical structure of dependencies $N(m_{22} = 0) \equiv N_0$ and $N(m_{22} = 1) \equiv N_1$ was studied.

The two-dimensional array of the "orientation-phase" Mueller matrix element $m_{22}(p \times k)$ was scanned in the horizontal direction $x \equiv 1, \ldots, p$ with a step $\Delta x = 1\,pix$.

Within each local sample $\left(1_{pix} \times n_{pix}\right)^{(k=1,2,\ldots,m)}$, the number (N) of characteristic values was calculated $m_{22}(k) = 0, - \left(N_0^{(k)}\right)$ and $m_{22}(k) = 1, - \left(N_1^{(k)}\right)$.

Thus, the dependences $N_0(x) \equiv \left(N_0^{(1)}, N_0^{(2)}, \ldots, N_0^{(p)}\right)$ and $N_1(x) \equiv \left(N_1^{(1)}, N_1^{(2)}, \ldots, N_1^{(p)}\right)$ giving the number of extreme values of the "orientation" matrix element m_{22} were determined within its coordinate distribution $(p \times k)$.

This approach made it possible to separately study the statistical manifestations of birefringence of myosin fibrils of myocardium tissue sections at its two extreme levels—the minimum—optically isotropic ($N(m_{22} = 1) \equiv N_1$) and the maximum—optically anisotropic ($N(m_{22} = 0) \equiv N_0$).

In Fig. 3.2 the coordinate and quantitative dependences of the extreme values of the "orientation" element $m_{22}(p \times k)$ of the Mueller matrix of myocardium tissue samples with ACI are presented.

Figure 3.3 shows the coordinate and quantitative dependences of the extreme values of the "orientational" element $m_{22}(p \times k)$ of the Mueller matrix of CCHD myocardium tissue samples. As a result of studies of the dependences of the number of extreme values $N(m_{22} = 1) \equiv N_1$ and $(m_{22} = 0) \equiv N_0$, diagnostic sensitivity to verification of deaths as a result of ACI distributions $N(m_{22} = 0) \equiv N_0$ was revealed.

For ACI, there was a decrease in the number of extreme values $m_{22} = 0$ of the "orientation" element of the Mueller matrix of myocardium tissue (Fig. 3.3a, c). The

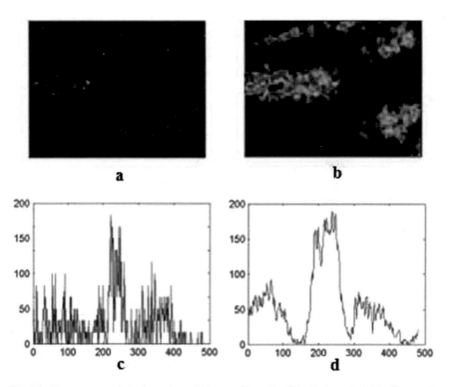

Fig. 3.2 The structure of the dependence $N(m_{22} = 0) \equiv N_0$ of the "orientation" Mueller matrix element m_{22} of the optic-anisotropic component of the myocardium tissue of both groups. Here **a** coordinate structure in ACI group; **b** coordinate structure in CCHD group; **c** quantitative structure in the ACI group; **d** quantitative structure in the CCHD group

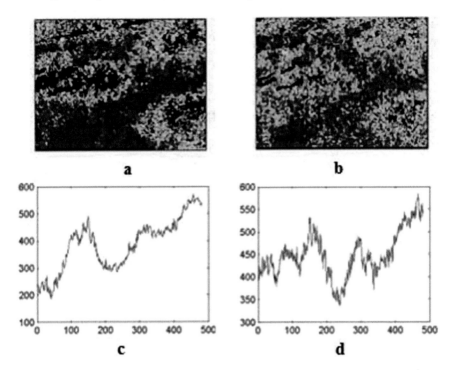

Fig. 3.3 The structure of the dependence $N(m_{22} = 1) \equiv N_1$ of the "orientation" Mueller matrix element m_{22} of the optically-anisotropic component of the myocardium tissue of both groups. Here **a** coordinate structure in ACI group; **b** coordinate structure in CCHD group; **c** quantitative structure in the ACI group; **d** quantitative structure in the CCHD group

indicated fact suggests the disordering of the directions of the optical axes, which are determined by the directions of the ends of optically-anisotropic myosin fibrils.

Quantitatively, the processes of changes in the birefringence of myosin myocardium fibrils at different extreme levels were characterised by a set of statistical moments $M_{j=1;2;3;4}(N_{0;1})$ that in turn characterised the distribution of the values of the dependences N_0 and N_1 (Table 3.2).

The obtained data on the statistical structure of the distribution of the number of extreme values of the "orientation" element m_{22} of the myocardial tissue Mueller matrix indicated the possibility of differentiating the cause of death and diagnosing ACI.

The most sensitive were the statistical moments of the 1st and 4th orders of distributions characterising the distribution $N_0(x)$ of the extreme values of the Mueller matrix element $m_{22}(p \times k) = 0$ of myocardium tissue. The following ranges of maximum differences between the statistical parameters characterising the distributions $m_{22}(p \times k) = 0$ describing the optically anisotropic component of the myocardium with ACI were established:

- average $M_1(N_0)$ (2.18 times increase);

Table 3.2 Statistical moments of the first to fourth orders of magnitude characterising the distribution $N_0(x)$ and $N_1(x)$ of extreme values of the "orientation" element of the Mueller matrix of the myocardium

Statistical moments	Cause of death		
	Control (n = 20)	CCHD (n = 69)	ACI (n = 69)
N_0			
Average, M_1	0.37 ± 0.021	0.61 ± 0.052	0.53 ± 0.046
P_1		<0.001	<0.001
P_2		<0.001	
Dispersion, M_2	0.21 ± 0.019	0.14 ± 0.012	0.17 ± 0.014
P_1		<0.001	<0.001
P_2		>0.05	
Skewness, M_3	0.87 ± 0.069	0.65 ± 0.056	0.74 ± 0.063
P_1		<0.001	<0.001
P_2		>0.05	
Kurtosis, M_4	0.68 ± 0.051	0.42 ± 0.037	0.55 ± 0.046
P_1		<0.001	<0.001
P_2		<0.001	
N_1			
Average, M_1	0.59 ± 0.039	0.53 ± 0.038	0.56 ± 0.037
P_1		>0.05	>0.05
P_2		>0.05	
Dispersion, M_2	0.19 ± 0.016	0.15 ± 0.012	0.17 ± 0.015
P_1		>0.05	>0.05
P_2		>0.05	
Skewness, M_3	0.75 ± 0.061	0.68 ± 0.054	0.79 ± 0.058
P_1		>0.05	>0.05
P_2		>0.05	
Kurtosis, M_4	0.86 ± 0.061	0.74 ± 0.063	0.78 ± 0.075
P_1		<0.001	<0.001
P_2		>0.05	

- dispersion $M_2(N_1)$ (2.3 times increase);
- skewness $M_3(N_0)$ (3.49 times increase);
- kurtosis $M_4(N_0)$ (2.88 times increase).

However, such a spread is caused not only by changes in the orientational structure of the morphological structure of the myocardium.

A large contribution is made by the "uncontrolled" angular positions of the samples relative to the direction of the irradiating beam. Therefore, a significant level of ambiguity arises in the differentiation of cases of CCHD and ACI, which certainly reduces the effectiveness of the method.

Correlation and Fractal Analysis

The results of experimental studies of the coordinate $m_{22}(p \times k)$ and statistical $N(m_{22})$ correlation $K(m_{22})$ and fractal $\log L(m_{22}) - \log {}^1\!/_d$ structures of the "orientation" elements m_{22} of the elements of the human myocardium Mueller matrix with CCHD and ACI are illustrated in Fig. 3.4.

An analysis of the figure shows that the coordinate distributions $m_{22}(p \times k)$ (Fig. 3.4a, b) of myocardium tissue with CCHD or ACI were multifractal, as evidenced by a monotonic decrease in the autocorrelation functions $K(m_{22})$ (Fig. 3.4c, d) and the presence of two stable slopes of the approximating curves to the dependences $\log L(m_{22}) - \log {}^1\!/_d$.

Quantitatively, the correlation structure of the distributions of the values of the "orientational" matrix elements $m_{22}(p \times k)$, defined for myocardium samples of the studied groups, is illustrated by the averaged values and ranges of changes in the

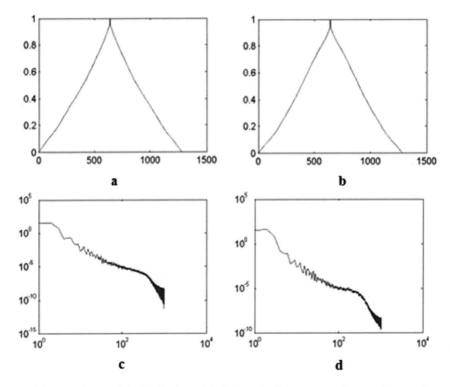

Fig. 3.4 Dependences of the distributions of the "orientation" element m_{22} of myocardium sections of both groups. Here **a** correlation dependences of distributions in a group with CCHD; **b** correlation dependencies of distributions in a group with ACI; **c** fractal dependencies of distributions in a group with CCHD; **d** fractal dependencies of distributions in the group with ACI

Table 3.3 Statistical moments $M_j(m_{22})$ of coordinate distributions $m_{22}(p \times k)$ of myocardium tissue

Statistical moments	Cause of death		
	Control (n = 20)	CCHD (n = 69)	ACI (n = 69)
m_{22}			
Dispersion, $K_2(m_{22})$	0.29 ± 0.021	0.32 ± 0.024	0.33 ± 0.024
P_1		>0.05	>0.05
P_2		>0.05	
Kurtosis, $K_4(m_{22})$	0.2 ± 0.012	0.24 ± 0.015	0.21 ± 0.013
P_1		>0.05	>0.05
P_2		>0.05	

2nd and 4th order correlation moments—dispersion $K_2(m_{22})$ and kurtosis $K_4(m_{22})$ (Table 3.3).

A comparative analysis of the averaged values and ranges of changes in the values of the correlation parameters $K_2(m_{22})$, $K_4(m_{22})$ that characterised the dependences of the autocorrelation distribution functions of the "orientation" elements m_{22} of the myocardium Mueller matrix did not reveal an objective possibility of verifying the cause of death, since the ranges of changes in the values of correlation parameters for different myocardium groups overlapped.

In order to search for more sensitive diagnostic criteria, we studied the statistical structure of the dependencies $N(m_{22} = 0) \equiv N_0$ and $N(m_{22} = 1) \equiv N_1$ (Figs. 3.5 and 3.6).

The results of a comparative study of the averaged values and ranges of variation of dispersion $K_2(N_0)$, $K_2(N_1)$, and kurtosis $K_4(N_0)$, $K_4(N_1)$, characterising the distributions $N(m_{22} = 0) \equiv N_0$, $N(m_{22} = 1) \equiv N_1$ of the number of extreme values $m_{22} = 0$ and $m_{22} = 1$ of the "orientation" element $m_{22}(p \times k)$ of the Mueller matrix of myocardium tissue with CCHD and ACI are presented in Table 3.4 ($N(m_{22} = 0) \equiv N_0$) and Table 3.5 ($N(m_{22} = 1) \equiv N_1$).

From the obtained experimental data on the statistical structure of the distribution of the number of extreme values of the phase element of the Mueller matrix of sections of myocardium tissue of both types, there is an objective possibility of diagnosing ACI and differentiating the cause of death. The kurtosis undergoes significant changes, which characterises the severity of the peak of the autocorrelation distribution function $K(N_0)$, for which the value $K_4(N_0)$ over the entire statistical sample of myocardium section samples increases 2.33 times.

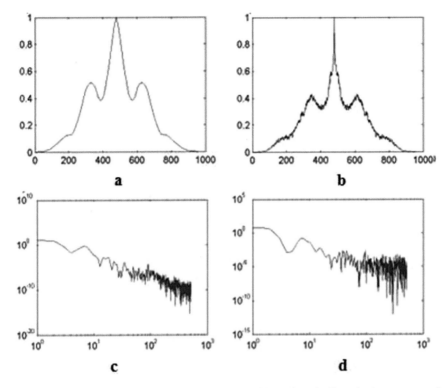

Fig. 3.5 Structure of the dependence of extreme values of the "orientation" matrix element $m_{22} = 0$ of myocardium sections from both groups. Here **a** is the correlation structure in the group with CCHD; **b** correlation structure in the group with ACI; **c** fractal structure in the group with CCHD; **d** fractal structure in the group with ACI

3.2   Diagnostic Capabilities of a Multi-parameter Analysis of the Distribution of the "Orientation-Phase" Element of the Myocardium Tissue of the Mueller Matrix

The results of experimental studies of the "orientation-phase" elements of the myocardium Mueller matrix in CCHD and ACI are illustrated in Fig. 3.7.

It can be seen from the obtained data that a qualitative comparison of the coordinate structure of the "orientation-phase" element m_{34} of the myocardium Mueller matrix alone makes it difficult to establish the cause of death, because the distributions $m_{34}(p \times k)$ characterising the orientation of the optical axes of myosin myocardium fibrils are coordinate-heterogeneous and have the same range of eigenvalues $-1 \le m_{34} \le 1$.

Quantitatively, the distribution structure of the "orientation-phase" elements m_{34} determined for the studied groups of myocardium samples was illustrated by the

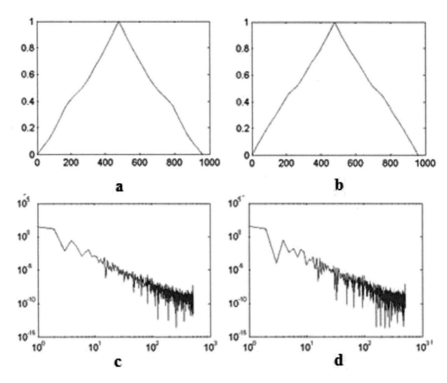

Fig. 3.6 Structure of the dependence of extreme values of the "orientation" matrix element $m_{22} = 1$ of myocardium sections deceased of both groups. Here **a** is the correlation structure in the group with CCHD; **b** correlation structure in the group with ACI; **c** fractal structure in the group with CCHD; **d** fractal structure in the group with ACI

Table 3.4 Correlation and fractal parameters of the dependences of the number of extreme values $N_0(m_{22} = 0)$ of the coordinate distributions $m_{22}(p \times k)$ of myocardium tissue

Statistical moments	Cause of death		
	Control (n = 20)	CCHD (n = 69)	ACI (n = 69)
N_0			
Dispersion, $K_2(N_0)$	0.05 ± 0.003	0.12 ± 0.011	0.09 ± 0.007
P_1		<0.001	<0.001
P_2		>0.05	
Kurtosis, $K_4(N_0)$	4.21 ± 0.36	2.34 ± 0.21	3.31 ± 0.27
P_1		<0.001	<0.001
P_2		<0.001	

Table 3.5 Correlation and fractal parameters of the dependences of the number of extreme values $N_1(m_{22} = 1)$ of the coordinate distributions $m_{22}(p \times k)$ of myocardium tissue

Statistical moments	Cause of death		
	Control (n = 20)	CCHD (n = 69)	ACI (n = 69)
N_1			
Dispersion, $K_2(N_1)$	0.25 ± 0.02	0.22 ± 0.018	0.24 ± 0.019
P_1		>0.05	>0.05
P_2		>0.05	
Kurtosis, $K_4(N_1)$	0.1 ± 0.009	0.11 ± 0.009	0.1 ± 0.009
P_1		>0.05	>0.05
P_2		>0.05	

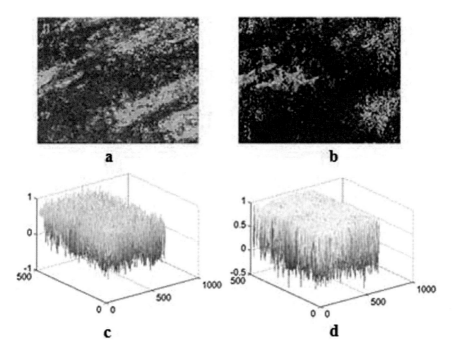

Fig. 3.7 Coordinate and statistical structure of the "orientation-phase" matrix element $m_{34}(p \times k)$ of myocardium of both groups. Here **a** is the coordinate structure in the group with CCHD; **b** coordinate structure in a group with ACI; **c** statistical structure in the group with CCHD; **d** statistical structure in the group with ACI

averaged values and ranges of variation of the values of statistical moments of the 1st–4th orders $M_{j=1;2;3;4}(m_{34})$ (Table 3.6).

A comparative analysis of the magnitudes and ranges of changes in the values of statistical moments of 1st–4th orders $M_{j=1;2;3;4}(m_{34})$ characterised the coordinate

Table 3.6 Statistical moments $M_{j=1;2;3;4}(m_{34})$ of coordinate distributions $m_{34}(p \times k)$ of myocardium tissue

Statistical moments	Cause of death		
	Control (n = 20)	CCHD (n = 69)	ACI (n = 69)
Average, $M_1(m_{34})$	0.14 ± 0.012	0.11 ± 0.011	0.13 ± 0.012
P_1		>0.05	>0.05
P_2		>0.05	
Dispersion, $M_2(m_{34})$	0.27 ± 0.023	0.32 ± 0.031	0.29 ± 0.027
P_1		>0.05	>0.05
P_2		>0.05	
Skewness, $M_3(m_{34})$	0.59 ± 0.054	0.54 ± 0.046	0.56 ± 0.053
P_1		>0.05	>0.05
P_2		>0.05	
Kurtosis, $M_4(m_{34})$	0.94 ± 0.087	0.87 ± 0.079	0.92 ± 0.87
P_1		>0.05	>0.05
P_2		>0.05	

distributions of myocardium phase elements within the studied groups and did not reveal an objective possibility of differentiating such cases of death, since the ranges of average $M_1(m_{34})$, dispersion $M_2(m_{34})$, skewness $M_3(m_{34})$ and kurtosis $M_4(m_{34})$ overlapped for different study groups.

In order to search for more sensitive diagnostic criteria, a statistical structure of parametric dependencies is established. $N(m_{34} = 0) \equiv N_0$ and $N(m_{34} = 1) \equiv N_1$.

Figures 3.8 and 3.9 show the coordinate and quantitative dependences of the extreme values of the "orientation-phase" element $m_{34}(p \times k)$ of the Mueller matrix of myocardium tissue samples with ACI and CCHD.

As a result of studies of the dependences of the number of extreme values $N(m_{34} = 1) \equiv N_1$ and $N(m_{34} = 0) \equiv N_0$, a diagnostic sensitivity to the diagnosis of ACI and differentiation of deaths as a result of CCHD and ACI distributions $N(m_{34} = 1) \equiv N_1$ was revealed.

Quantitatively, the changes in birefringence of myosin myocardium fibrils at different extreme levels were characterised by a set of statistical moments $M_{j=1;2;3;4}(N_{0;1})$ characterising the dependences N_0 and N_1 (Table 3.7).

From the obtained experimental data of the statistical structure of the distribution of the number of extreme values of the "orientation-phase" element m_{34} of the Mueller matrix of the myocardium tissue of the studied groups, there is an objective possibility of diagnosing ACI and differentiating the cause of death.

The most informative parameters were the statistical moments of the 3rd and 4th orders of the distribution $N_1(x)$ of extreme values of the Mueller matrix element $m_{34}(p \times k) = 1$ of myocardium tissue.

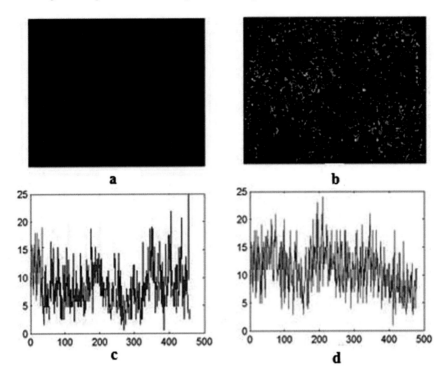

Fig. 3.8 Coordinate and quantitative structure of the dependences $N(m_{34} = 1) \equiv N_1$ of the "orientation-phase" element m_{34} of the optic-anisotropic component of the myocardium tissue of both groups. Here **a** is the coordinate structure in the group with CCHD; **b** coordinate structure in a group with ACI; **c** quantitative structure in the group with CCHD; **d** quantitative structure in the group with ACI

The following (maximum within the studied representative samples of myocardium sections) ranges of differences between the statistical parameters characterising the distributions $m_{34}(p \times k) = 1$ describing the optically anisotropic component of the myocardium with ACI and CCHD were established:

- skewness $M_3(N_1)$ (2.4 times increase);
- kurtosis $M_4(N_1)$ (2.2 times increase).

The results of studies of the correlation $K(m_{34})$ and fractal $\log L(m_{34}) - \log {}^1/_d$ structure of the "orientation-phase" elements m_{34} of the Mueller matrix of human myocardium with CCHD and ACI are illustrated in Fig. 3.10.

According to the data obtained, it can be seen that the coordinate distributions $m_{34}(p \times k)$ (0. a, b) of myocardium tissue with CCHD or ACI are multifractal, since there was a monotonic decrease in autocorrelation functions $K(m_{34})$ (Fig. 3.10, c, d) and the presence of two stable slopes of the approximating curves to the dependences $\log L(m_{34}) - \log {}^1/_d$.

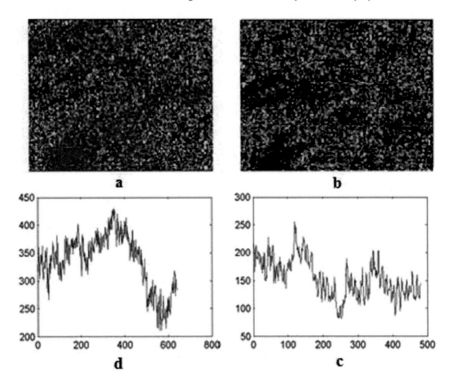

Fig. 3.9 Coordinate and quantitative structure of the dependences $N(m_{22} = 0) \equiv N_0$ of the "orientation-phase" element m_{34} of the optic-anisotropic component of the myocardium tissue of both groups. Here **a** is the coordinate structure in the group with CCHD; **b** coordinate structure in a group with ACI; **c** quantitative structure in the group with CCHD; **d** quantitative structure in the group with ACI

Quantitatively, the correlation structure of the distributions of the "orientational" matrix elements $m_{34}(p \times k)$ defined for two groups of myocardium samples is illustrated by the averaged values and ranges of variation of the dispersion $K_2(m_{34})$ and kurtosis $K_4(m_{34})$ (Table 3.8).

A comparative analysis of the values and ranges of changes in the values of correlation parameters $K_2(m_{34})$, $K_4(m_{34})$ autocorrelation distribution functions of the "orientation-phase" elements m_{34} of the Mueller matrix of myocardium sections within the studied groups did not reveal the objective possibility of setting ACI as the cause of death.

The ranges of changes in the values of correlation and fractal parameters for different groups overlapped.

In order to search for more sensitive diagnostic criteria, we studied the correlation and fractal structure of dependencies $N(m_{34} = 1) \equiv N_1$ (Fig. 3.11) and $N(m_{34} = 0) \equiv N_0$ (Fig. 3.12).

Table 3.7 Statistical moments of the 1st– 4th orders of distributions $N_0(x)$ and $N_1(x)$ of the extreme values of the "orientation-phase" element m_{34} of the Mueller matrix of myocardial tissue

Statistical moments	Cause of death		
	Control (n = 20)	CCHD (n = 69)	ACI (n = 69)
N_0			
Average, $M_1(N_0)$	0.53 ± 0.045	0.54 ± 0.046	0.51 ± 0.049
P_1		>0.05	>0.05
P_2		>0.05	
Dispersion, $M_2(N_0)$	0.28 ± 0.021	0.21 ± 0.017	0.23 ± 0.022
P_1		<0.001	<0.001
P_2		>0.05	
Skewness, $M_3(N_0)$	1.87 ± 0.16	1.46 ± 0.13	1.58 ± 0.15
P_1		<0.001	<0.001
P_2		>0.05	
Kurtosis, $M_4(N_0)$	1.16 ± 0.11	1.09 ± 0.095	1.12 ± 0.11
P_1		>0.05	>0.05
P_2		>0.05	
N_1			
Average, $M_1(N_1)$	0.12 ± 0.01	0.09 ± 0.0086	0.11 ± 0.009
P_1		>0.05	>0.05
P_2		>0.05	
Dispersion, $M_2(N_1)$	0.39 ± 0.026	0.45 ± 0.034	0.41 ± 0.032
P_1		>0.05	>0.05
P_2		>0.05	
Skewness, $M_3(N_1)$	3.75 ± 0.28	2.11 ± 0.17	3.08 ± 0.26
P_1		<0.001	<0.001
P_2		<0.001	
Kurtosis, $M_4(N_1)$	2.86 ± 0.21	1.84 ± 0.18	2.37 ± 0.22
P_1		<0.001	<0.001
P_2		<0.001	

The results of a comparative study of the averaged values and ranges of variation of dispersion $K_2(N_0)$, $K_2(N_1)$; of kurtosis $K_4(N_0)$, $K_4(N_1)$ of distributions $N(m_{34} = 0) \equiv N_0$, $N(m_{34} = 1) \equiv N_1$, of the number of extreme values $m_{34} = 0$ and $m_{34} = 1$ of the "orientation-phase" element $m_{34}(p \times k)$ of the Mueller matrix of myocardial tissue with CCHD and ACI are presented in Table 3.9 ($N(m_{34} = 1) \equiv N_1$) and Table 3.10 ($N(m_{34} = 0) \equiv N_0$).

The results of studies of the statistical structure of the distributions of the number of extreme values of the phase element of the Mueller matrix of myocardium tissue indicate the possibility of establishing ACI and CCHD differentiation based on measuring

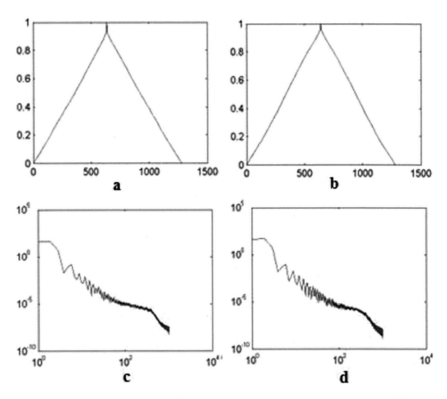

Fig. 3.10 Correlation and fractal parameters characterising the distribution of the "orientation-phase" m_{34} element of myocardium. Here **a** correlation parameters in the group with CCHD; **b** correlation parameters in the group with ACI; **c** fractal parameters in the group with CCHD; **d** fractal parameters in the group with ACI

Table 3.8 Correlation moments of autocorrelation functions of distribution $m_{34}(p \times k)$ of myocardium tissue

Statistical moments	Cause of death		
	Control (n = 20)	CCHD (n = 69)	ACI (n = 69)
Dispersion, $K_2(m_{34})$	0.029 ± 0.0021	0.31 ± 0.024	0.34 ± 0.026
P_1		>0.05	>0.05
P_2		>0.05	
Kurtosis, $K_4(m_{34})$	0.86 ± 0.077	0.84 ± 0.075	0.91 ± 0.079
P_1		>0.05	>0.05
P_2		>0.05	

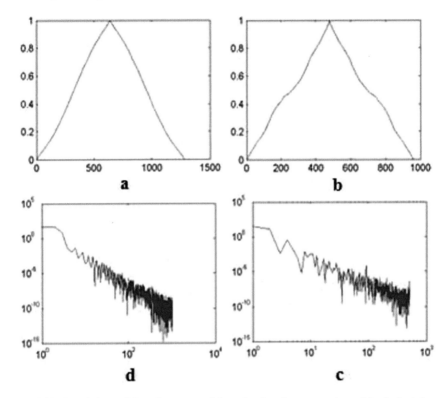

Fig. 3.11 Correlation and fractal structure of dependencies of extreme values of the "orientation-phase" matrix element $m_{34} = 0$ of myocardium of both groups. Here **a** is the correlation structure in the group with CCHD; **b** correlation structure in the group with ACI; **c** fractal structure in the group with CCHD; **d** fractal structure in the group with ACI

the kurtosis of the autocorrelation distribution function $K_4(N_1)$. ACI is characterised by a decrease $K_4(N_1)$ of 1.85 times.

The results of the analysis of the informativeness of the methods are illustrated by the values of the operational characteristics characterising the diagnostic strength of Mueller matrix microscopy of myocardium sections (Tables 3.11 and 3.12).

An analysis of the operational characteristics of the parametric method of Mueller matrix microscopy, determining its diagnostic informativeness, did not reveal a sufficiently high level of balanced accuracy of most of the objective parameters characterising the Mueller matrix images of the orientation and orientation-phase structure of the myocardium fibrillar network in the task of its post-mortem diagnosis.

The following level of operational performance has been achieved:

$$m_{22} \Leftrightarrow \begin{cases} Se = 67\% - 68\% \\ Sp = 57\% - 58\% \\ Ac = 61\% - 62.5\% \end{cases} ; m_{34} \Leftrightarrow \begin{cases} Se = 61\% - 64\% \\ Sp = 55\% - 58\% \\ Ac = 58\% - 61\% \end{cases}.$$

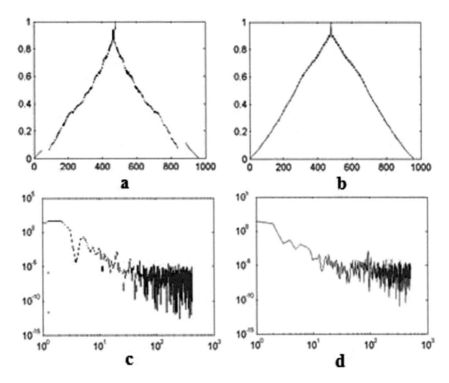

Fig. 3.12 Correlation and fractal structure of dependencies of extreme values of the "orientation-phase" matrix element $m_{34} = 1$ of myocardium of both groups. Here **a** is the correlation structure in the group with CCHD; **b** correlation structure in the group with ACI; **c** fractal structure in the group with CCHD; **d** fractal structure in the group with ACI

Table 3.9 Correlation parameters of the dependences of the number of extreme values $N_1(m_{34} = 0)$ of the coordinate distributions $m_{34}(p \times k)$ of myocardium tissue

Statistical moments	Cause of death		
	Control (n = 20)	CCHD (n = 69)	ACI (n = 69)
N_0			
Average, $K_2(N_0)$	0.23 ± 0.021	0.21 ± 0.019	0.24 ± 0.022
P_1		>0.05	>0.05
P_2		>0.05	
Dispersion, $K_4(N_0)$	0.37 ± 0.032	0.41 ± 0.035	0.39 ± 0.034
P_1		>0.05	>0.05
P_2		>0.05	

Table 3.10 Correlation parameters of the dependences of the number of extreme values $N_0(m_{34} = 1)$ of the coordinate distributions $m_{34}(p \times k)$ of myocardium tissue

Statistical moments	Cause of death		
	Control (n = 20)	CCHD (n = 69)	ACI (n = 69)
N_1			
Dispersion, $K_2(N_1)$	0.19 ± 0.017	0.14 ± 0.012	0.16 ± 0.013
P_1		<0.001	<0.001
P_2		>0.05	
Kurtosis, $K_4(N_1)$	1.21 ± 0.11	2.34 ± 0.21	1.53 ± 0.14
P_1		<0.001	<0.001
P_2		<0.001	

Table 3.11 Operational characteristics of a parametric analysis of the distributions of the orientational matrix element m_{22} of myocardium sections

R	$Se(\alpha)$, % "1 − 2"	$Sp(\alpha)$, % "1 − 3"	$Se(\alpha)$, %"2 − 3"	$Sp(\alpha)$, %"3 − 2"
Average, $M_1(N_0)$	67	57	67 $a = 46, b = 23$	57 $c = 39, d = 30$
Dispersion, $M_2(N_0)$	68 $a = 46, b = 23$	61 $c = 42, d = 27$	61 $a = 42, b = 27$	55 $c = 38, d = 31$
Skewness, $M_3(N)$	61 $a = 42, b = 27$	57 $c = 39, d = 30$	61 $a = 42, b = 27$	54 $c = 37, d = 32$
Kurtosis, $M_4(N_0)$	64 $a = 48, b = 21$	62 $c = 43, d = 26$	64 $a = 44, b = 25$	58 $c = 40, d = 29$
Kurtosis, $K(N_0)$	68 $a = 47, b = 22$	60 $c = 41, d = 28$	68 $a = 46, b = 23$	57 $c = 39, d = 30$

Table 3.12 Operational characteristics of a parametric analysis of the distributions of the orientation-phase m_{34} matrix element of myocardium sections

R	$Se(\alpha)$, %"1 − 2"	$Sp(\alpha)$, %"1 − 3"	$Se(\alpha)$, %"2 − 3"	$Sp(\alpha)$, %"3 − 2"
Skewness, $M_3(N_1)$	66 $a = 45, b = 24$	58 $c = 40, d = 29$	61 $a = 42, b = 27$	55 $c = 38, d = 31$
Kurtosis, $M_4(N_1)$	68 $a = 46, b = 23$	59 $c = 41, d = 28$	61 $a = 42, b = 27$	54 $c = 37, d = 32$
Kurtosis, $K_4(N_1)$	66 $a = 45, b = 24$	58 $c = 40, d = 29$	64 $a = 42, b = 27$	58 $c = 40, d = 29$

The achieved level of information content is slightly higher (by 5–7%) than the information content of direct polarisation mapping of microscopic images of myocardium sections.

However, the level of balanced accuracy does not exceed the satisfactory quality of the diagnostic test.

Such a result is also associated with the azimuthal dependence of the magnitude of the matrix elements m_{22} and m_{34} during rotation of the plane of the sample relative to the direction of irradiation. Indeed, the uniform orientation of myofibrils of myocardium samples relative to the plane of polarisation of the laser beam is impossible. Accordingly, the coordinate distributions of the corresponding elements of the Mueller matrix of various samples within the same group depend not only on the orientational structure of the fibrillar networks, but also on the specific location of the sample. Due to this, in each group, the average value of statistical and correlation moments decreases, and the range of variation of random values, on the contrary, increases. In accordance with this, the overall level of sensitivity, specificity and balanced accuracy of the method is reduced.

The azimuthal dependence of the Mueller matrix images of polycrystalline fibrillar networks of the myocardium section obtained for the rotation of the sample relative to the direction of the irradiating laser beam $\Theta = 0°$ and $\Theta = 45°$ is illustrated in Fig. 3.13.

Analysis of topographic $m_{22}(p \times k)$, $m_{34}(p \times k)$ and statistical structure $N(m_{22})$; $N(m_{34})$ (Fig. 3.13) of various ($\Theta = 0°$, $\Theta = 45°$) Mueller matrix images of a

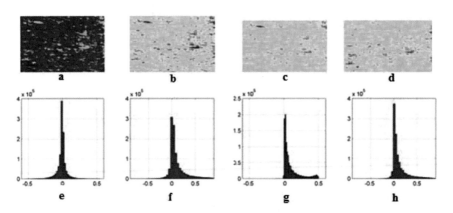

Fig. 3.13 Mueller matrix images $m_{22}(p \times k)$, $m_{34}(p \times k)$ and histograms of distributions $N(m_{22})$; $N(m_{34})$ myocardium section obtained for sample rotation relative to the direction of the irradiating laser beam $\Theta = 0°$ and $\Theta = 45°$. Here **a** is the topographic structure $m_{22}(p \times k)$ of the Mueller matrix image obtained to rotate the sample relative to the direction of the irradiating laser beam $\Theta = 0°$; **b** topographic structure $m_{34}(p \times k)$ of the Mueller matrix image obtained for rotation of the sample relative to the direction of the irradiating laser beam $\Theta = 0°$; **c** topographic structure $m_{22}(p \times k)$ of the Mueller matrix image obtained for rotation of the sample relative to the direction of the irradiating laser beam $\Theta = 45°$; **d** topographic structure $m_{34}(p \times k)$ of the Mueller matrix image obtained for rotation of the sample relative to the direction of the irradiating laser beam $\Theta = 45°$; **e** statistical structure $N(m_{22})$ of the Mueller matrix image obtained for rotation of the sample relative to the direction of the irradiating laser beam $\Theta = 0°$; **f** statistical structure $N(m_{34})$ of the Mueller matrix image obtained for rotation of the sample relative to the direction of the irradiating laser beam $\Theta = 0°$; **g** statistical structure $N(m_{22})$ of the Mueller matrix image obtained for rotation of the sample relative to the direction of the irradiating laser beam $\Theta = 45°$; **h**—statistical structure $N(m_{34})$ of the Mueller matrix image obtained for rotation of the sample relative to the direction of the irradiating laser beam $\Theta = 45°$

myocardium section confirmed the presence of their individual, heterogeneous, coordinate structures. The positions of the extremum, full width-half maximum, skewness, and sharpness of the peak of the dependences $N(m_{22})$ and $N(m_{34})$ are individual for different angles of rotation of the sample.

These data explain the differences between the values of statistical moments of the 1st–4th orders $M_{i=1,2,3,4}(\Theta)$, which characterise the coordinate distribution $m_{22}(p \times k)$, $m_{34}(p \times k)$.

In the theory of Mueller matrix biomedical optics, it is shown that only two elements of the matrix are azimuthally independent. The first is a phase element m_{44} associated with birefringence of the secondary protein structure—myosin fibrillar network. The second is the following combination of matrix elements $\Delta m = \frac{m_{23}-m_{32}}{m_{22}+m_{33}}$, which is associated with the optical activity of small-scale polypeptide chains of myosin molecules.

3.3 Diagnosis of Acute Myocardium Ischemia by Azimuthally Invariant Mueller Matrix Mapping

In this part of the work, the results of azimuthally invariant Mueller matrix mapping ($\{m_{44}(p \times k)\}(\Theta) = const$ and $\{\Delta m(p \times k)\}(\Theta) = const$) of myocardium sections are presented.

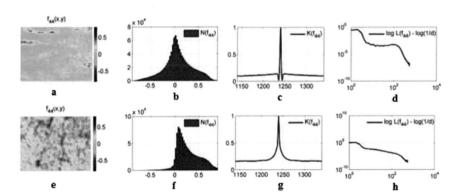

Fig. 3.14 Statistical $N(m_{44})$, correlation $K(m_{44})$ and spatial-frequency dependencies $\log L(m_{44}) - \log\left(\frac{1}{d}\right)$ characterising the Mueller matrix images $m_{44}(p \times k)$ of myocardium sections. Here **a** is the topographic structure m_{44} of the Muller-matrix image with CCHD; **b** statistical dependence $N(m_{44})$ with CCHD; **c** correlation dependence $K(m_{44})$ with CCHD; **d** spatial-frequency dependence $\log L(m_{44}) - \log\left(\frac{1}{d}\right)$ with CCHD; **e** topographic structure m_{44} of the Mueller matrix image with ACI; **f** statistical dependence $N(m_{44})$ with ACI; **g** correlation dependence $K(m_{44})$ with ACI; **h** spatial-frequency dependence $\log L(m_{44}) - \log\left(\frac{1}{d}\right)$ with ACI

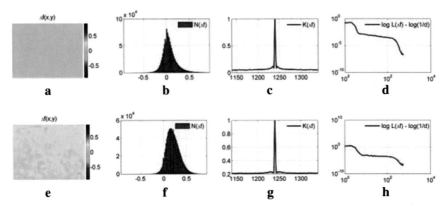

Fig. 3.15 Statistical $N(\Delta m)$, correlation $K(\Delta m)$ and spatial-frequency dependencies $\log L(\Delta m) - \log\left(\frac{1}{d}\right)$ characterising the Mueller matrix images $\Delta m(p \times k)$ of myocardium sections. Here **a** is the topographic structure Δm of the Muller-matrix image with CCHD; **b** statistical dependence $N(\Delta m)$ with CCHD; **c** correlation dependence $K(\Delta m)$ with CCHD; **d** spatial-frequency dependence $\log L(\Delta m) - \log\left(\frac{1}{d}\right)$ with CCHD; **e** topographic structure Δm of the Mueller matrix image with ACI; f—statistical dependence $N(\Delta m)$ with ACI; g—correlation dependence $K(\Delta m)$ with ACI; h—spatial-frequency dependence $\log L(\Delta m) - \log\left(\frac{1}{d}\right)$ with ACI

Figures 3.14 and 3.15 show the Mueller matrix images m_{44} (Fig. 3.14) and Δm (Fig. 3.15) of sections of the myocardium deceased due to CCHD and ACI.

The results of calculating the totality of objective statistical and correlation parameters of the azimuthally invariant Mueller matrix microscopy method for myocardium sections of the studied groups are given in Tables 3.13 and 3.14.

The following quantitative criteria have been established for the Mueller matrix images m_{44} of the sections of the group, allowing one to verify ACI:

$$m_{44}(p \times k)\begin{cases} \Delta M_1 = 1.26;\ \Delta M_2 = 1.27;\ \Delta M_3 = 1.53;\ \Delta M_4 = 1.56 \\ \Delta M_2^K = 1.22;\ \Delta M_4^K = 1.37 \end{cases}.$$

The following quantitative criteria have been established for the Mueller matrix images Δm of the sections of the group, allowing one to verify ACI:

$$\Delta m(p \times k)\begin{cases} \Delta M_1 = 1.29;\ \Delta M_2 = 1.27;\ \Delta M_3 = 1.57;\ \Delta M_4 = 1.38 \\ \Delta M_2^K = 1.29;\ \Delta M_4^K = 1.49 \end{cases}.$$

As can be seen from the above data, the statistical moments M_3, M_4 and $M^K{}_4$, characterising the coordinate distributions $m_{44}(p \times k)$, $\Delta m(p \times k)$, turned out to be the most sensitive ($\Delta \overline{q} = \max$).

The information analysis results give the values of operational characteristics characterising the diagnostic strength of Mueller matrix microscopy of myocardium sections (Tables 3.16 and 3.18).

Table 3.13 The average ($\overline{q}$) and standard deviation ($\pm\sigma$) of the Mueller matrix microscopy parameters r of the phase element m_{44} distributions of myocardium samples

Statistical moments	Cause of death		
	Control (n = 20)	CCHD (n = 69)	ACI (n = 69)
m_{44}			
Average, M_1	0.27 ± 0.021	0.45 ± 0.035	0.36 ± 0.034
P_1		<0.001	<0.001
P_2		<0.001	
Dispersion, M_2	0.18 ± 0.012	0.16 ± 0.014	0.19 ± 0.016
P_1		>0.05	>0.05
P_2		>0.05	
Skewness, M_3	1.31 ± 0.11	0.68 ± 0.058	1.14 ± 0.1
P_1		<0.001	<0.001
P_2		<0.001	
Kurtosis, M_4	1.18 ± 0.071	0.73 ± 0.069	0.97 ± 0.082
P_1		<0.001	<0.001
P_2		<0.001	
Dispersion, M^K_2	0.24 ± 0.019	0.22 ± 0.018	0.25 ± 0.023
P_1		>0.05	>0.05
P_2		>0.05	
Kurtosis, M^K_4	0.79 ± 0.064	1.44 ± 0.12	0.93 ± 0.081
P_1		<0.001	<0.001
P_2		<0.001	

An analysis of the operational characteristics of the Mueller matrix microscopy method, which determined diagnostic informativeness, revealed an increase in the level of balanced accuracy of the post-mortem diagnosis of myocardium ACI by cross-statistical analysis of the following objective parameters (statistical moments of the 3rd and 4th orders, correlation moment of the 4th order), which characterise the distribution of values of azimuthally independent Mueller matrix invariants describing the manifestations of optical anisotropy of fibrillar protein networks at the primary (Δm)) and secondary (m_{44}) levels:

$$m_{44} \Leftrightarrow \begin{cases} Se = 74\% - 75\% \\ Sp = 62\% - 64\% \\ Ac = 68.5\% - 69\% \end{cases};$$

$$\Delta m \Leftrightarrow \begin{cases} Se = 72\% - 74\% \\ Sp = 59\% - 62\% \\ Ac = 65.5\% - 68\% \end{cases}.$$

Table 3.14 The average ($\bar{q}$) and standard deviation ($\pm\sigma$) of the Mueller matrix microscopy parameters r of the phase element Δm distributions of myocardium samples

Statistical moments	Cause of death		
	Control (n = 20)	CCHD (n = 69)	ACI (n = 69)
Δm			
Average, M_1	0.08 ± 0.021	0.05 ± 0.0037	0.07 ± 0.0059
P_1		>0.05	>0.05
P_2		>0.05	
Dispersion, M_2	0.17 ± 0.012	0.13 ± 0.011	0.16 ± 0.013
P_1		>0.05	>0.05
P_2		>0.05	
Skewness, M_3	0.31 ± 0.021	0.17 ± 0.015	0.24 ± 0.023
P_1		<0.001	<0.001
P_2		<0.001	
Kurtosis, M_4	0.48 ± 0.031	0.29 ± 0.024	0.36 ± 0.034
P_1		<0.001	<0.001
P_2		<0.001	
Dispersion, M^K_2	0.09 ± 0.008	0.06 ± 0.0048	0.08 ± 0.0071
P_1		>0.05	>0.05
P_2		>0.05	
Kurtosis, M^K_4	2.57 ± 0.21	1.49 ± 0.11	2.13 ± 0.17
P_1		<0.001	<0.001
P_2		<0.001	

The results obtained (Tables 3.15 and 3.16) allow us to state the growth (by 10% -15%) of balanced accuracy compared to the direct Mueller matrix mapping method (Tables 3.11 and 3.12).

Table 3.15 Operational characteristics of multidimensional Mueller matrix microscopy of the distributions of the phase element m_{44} of myocardium sections

R	$Se(\alpha), \%\text{"1} - 2\text{"}$	$Sp(\alpha), \%\text{"1} - 3\text{"}$	$Se(\alpha), \%\text{"2} - 3\text{"}$	$Sp(\alpha), \%\text{"3} - 2\text{"}$
Skewness M_3	74 $a = 49, b = 20$	64 $c = 44, d = 25$	62 $a = 43, b = 26$	59 $c = 41, d = 28$
Kurtosis M_4	77 $a = 53, b = 16$	72 $c = 48, d = 21$	74 $a = 49, b = 20$	64 $c = 44, d = 25$
Kurtosis K_4	80 $a = 55, b = 14$	74 $c = 50, d = 19$	75 $a = 52, b = 17$	62 $c = 43, d = 26$

Table 3.16 Operational characteristics of multidimensional Mueller matrix microscopy of the distributions of the phase element Δm of myocardium sections

R	$Se(\alpha), \%\text{"}1 - 2\text{"}$	$Sp(\alpha), \%\text{"}1 - 3\text{"}$	$Se(\alpha), \%\text{"}2 - 3\text{"}$	$Sp(\alpha), \%\text{"}3 - 2\text{"}$
Skewness M_3	60 $a = 42, b = 27$	58 $c = 40, d = 29$	58 $a = 40, b = 29$	55 $c = 38, d = 31$
Kurtosis M_4	81 $a = 56, b = 13$	72 $c = 50, d = 19$	74 $a = 49, b = 23$	62 $c = 43, d = 26$
Kurtosis K_4	86 $a = 59, b = 10$	78 $c = 54, d = 15$	72 $a = 50, b = 19$	59 $c = 41, d = 28$

At the same time, a fairly high level of balanced accuracy is achieved, which from the standpoint of evidence-based medicine corresponds to a good quality diagnostic test—$Ac(m_{44}) = 69\%$ and $Ac(\Delta m) = 68\%$ with good reproducibility of experimental data.

As already noted above, the "information load" of Mueller matrix invariants m_{44} and Δm is different.

The first of them (m_{44}) mainly describes the manifestations of optical anisotropy at a large-scale level of myocardium fibrillar myosin networks. The second (Δm) is associated with small-scale levels of optically active myosin molecules (polypeptide chains). Therefore, in order to deepen the analysis of the experimental data of the azimuthally independent Mueller matrix mapping of myocardium sections of the studied groups, the information capabilities of the wavelet-analysis were established.

3.4 Post-mortem Diagnosis of the Myocardium Using Wavelet Analysis of Azimuthally Invariant Mueller Matrix Images of Sections of the Studied Groups

This section presents the results of studies of myocardium samples, using wavelet-analysis of the coordinate distributions of azimuthally independent elements of the Mueller matrix, characterising the primary and secondary structure of the proteins of the heart muscle in order to increase the balanced accuracy of the diagnostic method of post-mortem ACI diagnosis.

As noted, wavelet-analysis allows one to separately assess the manifestations of the optical properties of small-scale polypeptide chains of optically active myosin molecules (Mueller matrix images of the invariant Δm) and large-scale fibrillar myosin networks (Mueller matrix images m_{44}) of myocardium sections.

If we analyse the occurrence of ACI as a sequence of biochemical processes due to a shortage of the macroergic compound ATP and, accordingly, disruption of the actinomyosin complex, as the most energy-intensive process in cardiomyocytes, we can assume that the greatest differences in ACI should be sought on small scales of changes in the primary structure of myocardium protein polypeptide chains.

In contrast, the main optical manifestations of CCHD are in the transformation of large-scale fibrillar myosin networks. Therefore, a scale-selective approach to the analysis of Mueller matrix images of optically anisotropic structures of sections of both groups will increase the accuracy of the post-mortem diagnosis of myocardium ACI and provide differentiation from other states (CCHD).

The results of studies of the distributions of wavelet-coefficients characterising the manifestations of the optical properties of anisotropic myocardium networks at different scales of Mueller matrix images of sections of both groups are shown in Figs. 3.16 3.17, 3.18 and 3.19. These figures illustrate:

- Mueller matrix images $\{m_{44}(x, y); \Delta m(x, y)\}$ of myocardium samples, in the case of ACI and CCHD;
- two-dimensional distributions of wavelet-coefficients $W_{a,b}\{m_{44}(x, y); \Delta m(x, y)\}$ characterising the coordinate structure of the distributions of Mueller matrix

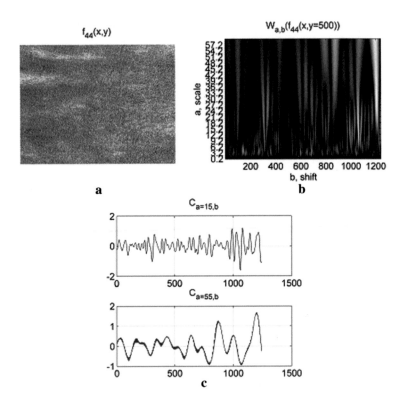

Fig. 3.16 Two-dimensional distributions of the Mueller matrix image m_{44}, wavelet-coefficients $W_{a,b}$ and their different-scale sections $C_{a=15,b}$ and $C_{a=55,b}$ for the cut of the myocardium with ACI. Here **a** is the two-dimensional distribution of the Mueller matrix image m_{44}; **b** two-dimensional distribution of wavelet-coefficients $W_{a,b}(m_{44}(x, y))$; **c** sections of wavelet-coefficients on the scales $C_{a=15,b}$ and $C_{a=55,b}$

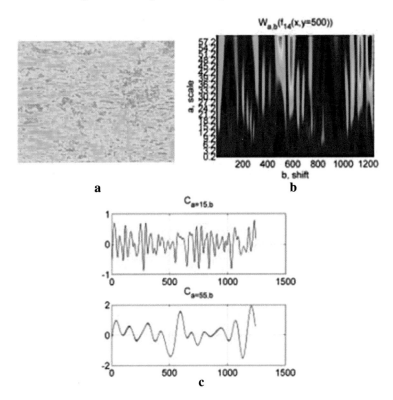

Fig. 3.17 Two-dimensional distributions of the Mueller matrix image m_{44}, wavelet-coefficients $W_{a,b}$ and their different-scale sections $C_{a=15,b}$ and $C_{a=55,b}$ for the cut of the myocardium with CCHD. Here **a** is the two-dimensional distribution of the Mueller matrix image m_{44}; **b** two-dimensional distribution of wavelet-coefficients $W_{a,b}(m_{44}(x, y))$; **c** sections of wavelet-coefficients on the scales $C_{a=15,b}$ and $C_{a=55,b}$

invariants, which is formed by different-scale morphological components of myosin fibrillar networks of the myocardium section;

- linear sections $C_{a=15,b}$ and $C_{a=55,b}$ of the wavelet-maps $W_{a,b}$ on the scale of the wavelet-function $a = 15$ and $a = 55$, which provide a separate assessment of the manifestations of changes in optical anisotropy at the level of polypeptide chains ($a = 15$), as well as fibrillar networks ($a = 55$).

A comparative analysis of the results of studying the multiscale coordinate distributions of wavelet-coefficients $W_{a,b}(m_{44})$ and $W_{a,b}(\Delta m)$ characterising the Mueller matrix images of optically anisotropic fibrillar structures of myocardium tissue samples with ACI and CCHD at the primary and secondary levels revealed the greatest differences between them at the level of small scale a_{min} distributions of wavelet-coefficients (Figs. 3.17 and 3.19). Quantitatively, this was indicated by the modulation of the "small-scale" dependencies $C_{a=15,b}(\Delta m)$ of such wavelet-maps.

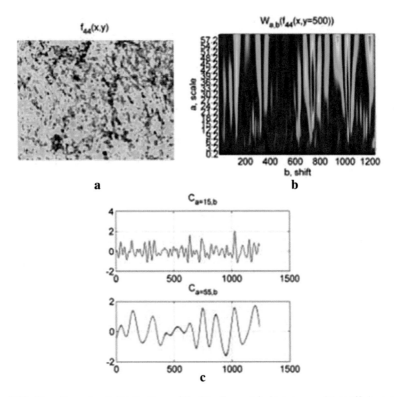

Fig. 3.18 Two-dimensional distributions of the Mueller matrix image, wavelet-coefficients $W_{a,b}$ and their different-scale sections $C_{a=15,b}$ and $C_{a=55,b}$ for the cut of the myocardium with ACI. Here **a** is the two-dimensional distribution of the Mueller matrix image Δm; **b** two-dimensional distribution of wavelet-coefficients $W_{a,b}(\Delta m(x, y))$; **c** sections of wavelet-coefficients on the scales $C_{a=15,b}$ and $C_{a=55,b}$

The most clearly defined changes were observed for the Mueller matrix image Δm, which characterised the manifestations of the primary protein structure of myosin networks of myocardium sections.

The revealed fact can be related to the fact that ischemic damage to polycrystalline myocardium fibrillar networks occurs not at the morphological (large-scale), but at the concentration (small-scale) levels of its structure. That is, this method visualises the formation of actinomyosin bridges at the level of the myosin heads and actin helix, which do not break as a result of a deficiency of macroergic compounds. Due to a change in the concentration of optically active myosin molecules as a resalt of which the occurrence of changes due to ACI, the level of optical activity decreased, which manifested itself in the formation of the coordinate distribution of the Mueller matrix invariant Δm of the corresponding section. Therefore, the depth of modulation of the amplitudes of the wavelet-coefficients at small scales $a_{\min}$ of this azimuthally-invariant Muller-matrix image of primary myosin structures decreased.

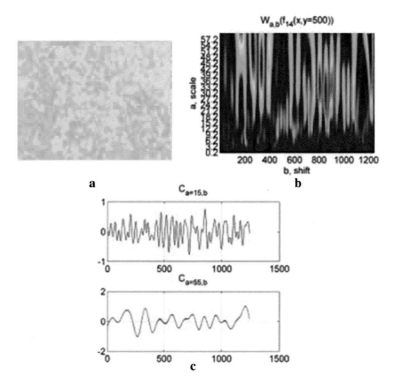

Fig. 3.19 Two-dimensional distributions of the Mueller matrix image, wavelet-coefficients $W_{a,b}$ and their different-scale sections $C_{a=15,b}$ and $C_{a=55,b}$ for the cut of the myocardium with CCHD. Here **a** is the two-dimensional distribution of the Mueller matrix image; **b** two-dimensional distribution of wavelet-coefficients $W_{a,b}(\Delta m(x, y))$; **c** sections of wavelet-coefficients on the scales $C_{a=15,b}$ and $C_{a=55,b}$

In the framework of the statistical approach, such a process was quantitatively characterised by a decrease in dispersion $M_{i=2}\left(C_{a=15,b}(\alpha)\right) \downarrow$ and a corresponding increase in the values of statistical moments of higher orders $M_{i=3;4}\left(C_{a=15,b}(\alpha)\right) \uparrow$ characterising the distribution of wavelet-coefficients—Table 3.17.

The most sensitive for the diagnosis of ACI turned out to be statistical moments M_2, M_3 and M_4 at the scale $a_{\min} = 15$ and M_3, M_4 at the scale $a_{\max}$.

Quantitative parameters of the difference between the statistical moments that characterise the distribution of wavelet-coefficients at different scales of the Mueller matrix map of the primary structure of the myosin network for statistically significant differentiation of myocardium samples deceased due to ACI from normal and CCHD are determined:

$$\Delta m(p \times k) \Rightarrow \begin{Bmatrix} a_{234\,\min} \\ a_{34\,\max} \end{Bmatrix}.$$

Table 3.17 Statistical moments of the first to fourth orders of magnitude characterising the distribution $C_{a=15,b}$ and $C_{a=55,b}$ wavelet-coefficients of the map of Mueller matrix images Δm of sections of myocardium tissues

Statistical moments	Cause of death		
	Control (n = 20)	CCHD (n = 69)	ACI (n = 69)
a_{min}			
Average, M_1	0.027 ± 0.0021	0.03 ± 0.0025	0.045 ± 0.0037
P_1		>0.05	<0.001
P_2		>0.05	
Dispersion, M_2	0.12 ± 0.009	0.23 ± 0.017	0.16 ± 0.013
P_1		<0.001	<0.001
P_2		<0.001	
Skewness, M_3	0.24 ± 0.019	0.55 ± 0.046	0.36 ± 0.034
P_1		<0.001	<0.001
P_2		<0.001	
Kurtosis, M_4	1.38 ± 0.11	0.81 ± 0.074	1.19 ± 0.11
P_1		<0.001	<0.001
P_2		<0.001	
$a_{max} = 55$			
Average, M_1	0.12 ± 0.009	0.09 ± 0.008	0.11 ± 0.009
P_1		>0.05	>0.05
P_2		>0.05	
Dispersion, M_2	0.23 ± 0.016	0.21 ± 0.015	0.22 ± 0.018
P_1		>0.05	>0.05
P_2		>0.05	
Skewness, M_3	0.58 ± 0.045	0.31 ± 0.028	0.44 ± 0.038
P_1		<0.001	<0.001
P_2		<0.001	
Kurtosis, M_4	0.86 ± 0.071	0.52 ± 0.048	0.68 ± 0.055
P_1		<0.001	<0.001
P_2		<0.001	

On the contrary, wavelet-analysis of Mueller matrix images m_{44} of the secondary structure of myosin grids of sections of both groups showed greater sensitivity to myocardium changes in ischemia on a large scale of the fibrillar network—Table 3.18. As can be seen, higher-order statistical moments that characterise the distribution of wavelet-coefficients on a large scale a_{max} of the Mueller matrix image $m_{44}(p \times k)$ turned out to be sensitive to ischemic changes at a large-scale morphological level of myocardium structure, namely M_3 and M_4. The quantitative values of differences

Table 3.18 Statistical moments of the 1st–4th orders characterising the distribution $C_{a=15,b}$ and $C_{a=55,b}$ wavelet-coefficients of the Mueller matrix images m_{44} of myocardium sections

Statistical moments	Cause of death		
	Control (n = 20)	CCHD (n = 69)	ACI (n = 69)
$a_{\min}$			
Average, M_1	0.022 ± 0.0021	0.023 ± 0.021	0.025 ± 0.0022
P_1		>0.05	>0.05
P_2		>0.05	
Dispersion, M_2	0.19 ± 0.014	0.18 ± 0.014	0.16 ± 0.013
P_1		>0.05	>0.05
P_2		>0.05	
Skewness, M_3	0.44 ± 0.039	0.45 ± 0.042	0.41 ± 0.034
P_1		>0.05	>0.05
P_2		>0.05	
Kurtosis, M_4	0.71 ± 0.065	0.67 ± 0.054	0.74 ± 0.069
P_1		>0.05	>0.05
P_2		>0.05	
$a_{\max}$			
Average, M_1	0.15 ± 0.011	0.14 ± 0.012	0.16 ± 0.013
P_1		>0.05	>0.05
P_2		>0.05	
Dispersion, M_2	0.29 ± 0.024	0.25 ± 0.021	0.28 ± 0.023
P_1		>0.05	>0.05
P_2		>0.05	
Skewness, M_3	1.21 ± 0.095	0.74 ± 0.068	0.96 ± 0.088
P_1		<0.001	<0.001
P_2		<0.001	
Kurtosis, M_4	1.46 ± 0.12	1.09 ± 0.094	1.31 ± 0.11
P_1		<0.001	<0.001
P_2		<0.001	

between such objective parameters for statistically significant differentiation of ACI of myocardium samples deceased from CCHD and the control group are determined:

$$m_{44}(p \times k) \Rightarrow \{a_{4\,\max}\}.$$

The information analysis results illustrate the values of operational characteristics characterising the diagnostic strength of the post-mortem diagnosis of myocardium ACI by wavelet-analysis of the Mueller matrix images of the heart muscle at the small ($a_{\min}$)—Tables 3.19 and 3.20 and large ($a_{\max}$)—Tables 3.21 and 3.22—scales.

Table 3.19 Operational characteristics of the wavelet-analysis of the distributions of Mueller matrix images Δm of myocardium sections on a scale a_{min}

R	$Se(\alpha)$, %"1 − 2"	$Sp(\alpha)$, %"1 − 3"	$Se(\alpha)$, %"2 − 3"	$Sp(\alpha)$, %"3 − 2"
Dispersion, M_2	85 $a = 58, b = 11$	74 $c = 51, d = 18$	77 $a = 53, b = 16$	68 $c = 47, d = 22$
Skewness, M_3	89 $a = 61, b = 9$	79 $c = 54, d = 21$	83 $a = 57, b = 12$	70 $c = 48, d = 21$
Kurtosis, M_4	94 $a = 64, b = 5$	87 $c = 60, d = 19$	86 $a = 59, b = 10$	72 $c = 50, d = 19$

Table 3.20 Operational characteristics of the wavelet-analysis of the distributions of Mueller matrix images m_{44} of myocardium sections on a scale a_{min}

R	$Se(\alpha)$, %"1 − 2"	$Sp(\alpha)$, %"1 − 3"	$Se(\alpha)$, %"2 − 3"	$Sp(\alpha)$, %"3 − 2"
Skewness, M_3	92 $a = 62, b = 7$	86 $c = 57, d = 12$	77 $a = 53, b = 16$	72 $c = 50, d = 19$
Kurtosis, M_4	94 $a = 64, b = 5$	90 $c = 61, d = 8$	78 $a = 54, b = 15$	72 $c = 50, d = 19$

Table 3.21 Operational characteristics of the wavelet-analysis of the distributions of Mueller matrix images Δm of myocardium sections on a scale a_{max}

R	$Se(\alpha)$, %"1 − 2"	$Sp(\alpha)$, %"1 − 3"	$Se(\alpha)$, %"2 − 3"	$Sp(\alpha)$, %"3 − 2"
Skewness, M_3	94 $a = 64, b = 5$	88 $c = 59, d = 12$	78 $a = 54, b = 15$	72 $c = 50, d = 19$
M_4	94 $a = 64, b = 5$	90 $c = 61, d = 8$	81 $a = 56, b = 13$	71 $c = 49, d = 20$

Table 3.22 Operational characteristics of the wavelet-analysis of the distributions of Mueller matrix images m_{44} of myocardium sections on a scale a_{max}

R	$Se(\alpha)$, %"1 − 2"	$Sp(\alpha)$, %"1 − 3"	$Se(\alpha)$, %"2 − 3"	$Sp(\alpha)$, %"3 − 2"
Skewness, M_3	94 $a = 64, b = 5$	90 $c = 61, d = 8$	82 $a = 57, b = 12$	73 $c = 51, d = 18$
Kurtosis, M_4	94 $a = 64, b = 5$	92 $c = 62, d = 7$	84 $a = 59, b = 10$	74 $c = 52, d = 17$

An analysis of the operational characteristics of the scale-selective method of wavelet analysis of Mueller matrix microscopy data, which determine its diagnostic informativeness, revealed an increase (in comparison with direct analysis of Mueller matrix images) of the balanced accuracy of post-mortem myocardium diagnosis.

On large-scale (a_{max}) estimates of the manifestations of the optical anisotropy of the secondary protein structures of fibrillar networks, the third and fourth order statistical moments characterising the skewness and sharpness of the peak distribution of the wavelet-coefficients of the Mueller matrix image Z_{44} of sections of the myocardium of both groups of the deceased turned out to be the most informative

$$m_{44} \Leftrightarrow \begin{cases} Se = 82\% - 84\% \\ Sp = 72\% - 74\% \\ Ac = 77\% - 79\% \end{cases}.$$

On a small scale (a_{min}) of evaluating the manifestations of the optical activity of the primary structure of myosin proteins, the most informative were the statistical moments of the 3rd and 4th orders characterising the distribution of the wavelet-coefficients of the Mueller matrix invariant інваріанту Δm of myocardium sections

$$\Delta m \Leftrightarrow \begin{cases} Se = 83\% - 86\% \\ Sp = 72\% - 74\% \\ Ac = 77.5\% - 79\% \end{cases}.$$

The obtained results on determining the operational characteristics of the polarisation microscopy method with scale-selective analysis (Tables 3.15 and 3.16) revealed an increase (up to 10%) of a good level of balanced accuracy of the diagnostic test - $Ac(m_{44}) = 79\%$ and $Ac(\Delta m) = 79\%$.

Thus, the presented method of post-mortem myocardium diagnosis based on the wavelet-analysis method of azimuthally-invariant Mueller matrix mapping of myocardium sections expands the functionality of the diagnostic test with high quality balanced accuracy and good data reproducibility.

3.5 Conclusions

1. Experimentally, within the framework of traditional Mueller matrix microscopy, the statistical moments (of the first to fourth orders of magnitude), correlation (autocorrelation functions) and fractal (logarithmic dependences of power spectra) distribution structures of the "orientational" and "orientational-phase" elements of the Mueller matrix fibrillar networks of myocardium sections of the studied groups were investigated. The azimuthal dependence of the coordinate distributions of the Mueller matrix images was revealed, which leads to poor reproducibility of the data and a rather low level of balanced accuracy of the post-mortem myocardium diagnosis.

2. When testing the method of azimuthally-invariant Mueller matrix mapping of post-mortem changes in the optical anisotropy of myocardium sections, a good level of balanced accuracy ($Ac \sim 72\%$) for detecting ACI has been obtained.

3. The possibilities of using wavelet-analysis of azimuthally-independent Mueller matrix invariants $r = \{m_{44}(x, y); \Delta m(x, y)\}$ of manifestations of optical anisotropy at the primary and secondary levels of the structure of myosin section networks in the post-mortem diagnosis of myocardium ACI are considered. The wavelet-analysis of the Mueller matrix images of the optical manifestations of the primary and secondary structure of the myocardium fibrillar network

revealed the sensitivity of a set of statistical moments of the first and fourth orders characterising the distribution of the amplitudes of the wavelet-coefficients at different scales of the morphological structure to post-mortem changes in the myocardium. On this basis, post-mortem diagnosis of myocardium ACI with greater balanced accuracy was implemented $Ac = 79\%$.

References

1. V. Tuchin, L. Wang, D. Zimnjakov, *Optical Polarisation in Biomedical Applications* (Springer, New York, USA, 2006)
2. Y. Ushenko, V. Ushenko, A. Dubolazov, V. Balanetskaya, N. Zabolotna, Mueller matrix diagnostics of optical properties of polycrystalline networks of human blood plasma. Opt. Spectrosc. **112**(6), 884–892 (2012)
3. V. Ushenko, O. Dubolazov, A. Karachevtsev, Two wavelength Mueller matrix reconstruction of blood plasma films polycrystalline structure in diagnostics of breast cancer. Appl. Opt. **53**(10), B128 (2016)
4. Y. Ushenko, G. Koval, A. Ushenko, O. Dubolazov, V. Ushenko, O. Novakovskaia, Mueller matrix of laser-induced autofluorescence of polycrystalline films of dried peritoneal fluid in diagnostics of endometriosis. J. Biomed. Opt. **21**(7), 071116 (2016)
5. R. Chipman, Polarimetry, *Handbook of Optics vol I— Geometrical and Physical Optics, Polarised Light, Components and Instruments*, ed. by M. Bass. (McGraw-Hill Professional, New York, 2010), pp. 22.1–22.37
6. N. Ghosh, M. Wood, A. Vitkin, Polarised light assessment of complex turbid media such as biological tissues via Mueller matrix decomposition, ed. by V. Tuchin, *Handbook of Photonics for Biomedical Science* (CRC Press, Taylor & Francis Group, London, 2010), pp. 253–282
7. S. Jacques, Polarised light imaging of biological tissues, in *Handbook of Biomedical Optics*. ed. by D. Boas, C. Pitris, N. Ramanujam (CRC Press, Boca Raton, London, New York, 2011), pp. 649–669
8. N. Ghosh, Tissue polarimetry: concepts, challenges, applications, and outlook. J. Biomed. Opt. **16**(11), 110801 (2011)
9. M. Swami, H. Patel, P. Gupta, Conversion of 3×3 Mueller matrix to 4×4 Mueller matrix for non-depolarising samples. Opt. Commun. **286**, 18–22 (2013)
10. D. Layden, N. Ghosh, A. Vitkin, Quantitative polarimetry for tissue characterisation and diagnosis, in *Advanced Biophotonics: Tissue Optical Sectioning. Boca Raton*, ed. by R. Wang, V. Tuchin (CRC Press, Taylor & Francis Group, London, New York, 2013), pp. 73–108
11. T. Vo-Dinh, *Biomedical Photonics Handbook: 3 volume Set*, 2nd ed. (Boca Raton: CRC Press, 2014)
12. A. Vitkin, N. Ghosh, A. Martino, Tissue polarimetry, in *Photonics: Scientific Foundations, Technology and Applications*, 4th edn., ed. by D. Andrews (Wiley, Hoboken, New Jersey, 2015), pp. 239–321
13. V. Tuchin, *Tissue Optics: Light Scattering Methods and Instruments for Medical Diagnosis*, 2nd edn. (SPIE Press, Bellingham, Washington, USA, 2007)
14. W. Bickel, W. Bailey, Stokes vectors, Mueller Matrices, and Polarised Scattered Light. Am. J. Phys. **53**(5), 468–478 (1985)
15. A. Doronin, C. Macdonald, I. Meglinski, Propagation of coherent polarised light in turbid highly scattering medium. J. Biomed. Opt. **19**(2), 025005 (2014)
16. A. Doronin, A. Radosevich, V. Backman, I. Meglinski, Two electric field Monte Carlo models of coherent backscattering of polarised light. J. Opt. Soc. Am. A **31**(11), 2394 (2014)

17. P.A. Gopinathan, G. Kokila, M. Jyothi, C. Ananjan, L. Pradeep, S.H. Nazir: Study of collagen birefringence in different grades of oral squamous cell carcinoma using picrosirius red and polarised light microscopy. Scientifica, 802980 (2015)
18. L. Rich, P. Whittaker, Collagen and picrosirius red staining: a polarised light assessment of fibrillar hue and spatial distriburuon. Braz. J. Morphol. Sci. (2005)
19. S. Bancelin, A. Nazac, B. Haj Ibrahim, et al., Determination of collagen fiber orientation in histological slides using Mueller microscopy and validation by second harmonic generation imaging. Opt. Exp. **22**(19), 22561–22574 (2014)
20. A. Ushenko, V. Pishak, *Laser Polarimetry of Biological tissue: Principles and Applications, in Handbook of Coherent-Domain Optical Methods: Biomedical Diagnostics,*, ed. by V. Tuchin (Environmental and Material Science, 2004), pp. 93–138
21. O. Angelsky, A. Ushenko, Y. Ushenko, V. Pishak, A. Peresunko, Statistical, correlation and topological approaches in diagnostics of the structure and physiological state of birefringent biological tissues. Handbook Photon. Biomed. Sci. (2010), pp. 283–322
22. Y. Ushenko, T. Boychuk, V. Bachynsky, O. Mincer, Diagnostics of structure and physiological state of birefringent biological tissues: statistical, correlation and topological approaches, in *Handbook of Coherent-Domain Optical Methods*, ed. by V. Tuchin (Springer Science+Business Media, 2013)

Conclusions

The monograph proposes and substantiates a set of new forensic methods and objective criteria for establishing acute coronary insufficiency, by means of new azimuthally-independent and scale-selective methods of polarisation and Mueller matrix mapping of polycrystalline fibrillar networks, in post-mortem diagnosis of the myocardium.

From the presented results, the following conclusions can be drawn:

1. Azimuthally independent Mueller matrix invariants characterising the manifestations of the primary and secondary polycrystalline structure of myosin networks were used for experimentally reproducing analysis and description of post-mortem changes in the myocardium. This made it possible to find the relationship between a set of statistical (statistical moments of the 1st–4th orders), correlation (correlation moments of the 2nd and 4th orders) and fractal (logarithmic dependences of power spectra) parameters characterising the distribution of the values of polarisation maps, Mueller matrix images, the intensity of autofluorescence of myocardium sections and their changes when measured due to chronic coronary heart disease and acute coronary insufficiency.

2. Quantitatively, azimuthal-invariant polarisation maps and polarisation ellipticities of microscopic images of histological sections of the myocardium—deceased due to acute coronary insufficiency—are characterised by an extension of the range of changes in the distribution of azimuth (polypeptide chains) and ellipticity (fibrillar networks). The most informative parameters were the statistical moments of the 3rd and 4th orders, as well as the correlation moments of the 4th order, characterising the polarisation maps of microscopic images of myocardium sections. A satisfactory level of balanced accuracy of the diagnostic test for the detection of acute coronary insufficiency using the interval value of the correlation moment of the 4th order of the autocorrelation function is obtained—$Ac(\alpha) = 68.5\%$.

3. To increase the balanced accuracy of post-mortem myocardium diagnosis, a scale-selective wavelet-analysis of polarisation maps of microscopic images

of fibrillar myosin networks was used. The results obtained for the separate detection of changes in the primary and secondary structure of myosin revealed an increase in balanced accuracy to the level of a good quality diagnostic test $(Ac(\beta) = 73\%))$ when using the azimuth kurtosis on the scale of a diagnostic criterion a_{max}.

4. The most sensitive parameters for post-mortem diagnosis of the myocardium have been established. On a large scale (a_{max}), the evaluation of the manifestations of optical anisotropy of fibrillar networks turned out to be the most informative using the third-order statistical moment, which characterises the skewness of the distribution of the wavelet-coefficients of the polarisation ellipticity map; on a small scale (a_{min}), evaluating the manifestations of the optical activity of the primary structure of myosin proteins, the third and fourth order statistical moments were the most informative, characterising the distribution of the wavelet-coefficients of the polarisation azimuth maps of microscopic images of histological sections of the myocardium deceased due to chronic coronary heart disease and acute coronary insufficiency.

5. It was determined that the processes of transformation of the structure of the human myocardium, which are caused by the dynamics of their changes after death, turn out to be in the growth of crystallisation of myocardium fibrillar networks, which is quantitatively determined by an increase in the average, skewness and kurtosis, characterising the coordinate distributions of the Mueller matrix invariants, which are analogues of the orientational properties of such networks and are morphologically related with the concentration of the substance of optically active myosin structures.

6. An analysis of the operational characteristics of the Mueller matrix microscopy method revealed a satisfactory level of balanced accuracy $(Ac(m_{44}) = 68.5\%))$ of the post-mortem diagnosis of myocardium ischemic changes by cross-statistical analysis of objective parameters (statistical moments of the 3rd and 4th order, correlation moment of the 4th order) characterising the distribution of azimuthal-independent values Mueller matrix invariants describing the manifestations of optical anisotropy of fibrillar protein networks at the primary and secondary levels.

7. Statistical analysis of the wavelet-coefficients of the distributions of the values of the Mueller matrix images of myocardium sections provided a good level of balanced accuracy in the diagnosis of acute coronary insufficiency by applying the diagnostic method of Mueller matrix microscopy with scale-selective analysis - $Ac(m_{44}) = 79\%$ and $Ac(\Delta m) = 79\%$.

Printed in the United States
by Baker & Taylor Publisher Services